Springer Texts in Statistics

Advisors:
Stephen Fienberg   Ingram Olkin

Springer Texts in Statistics

Probability and Statistical Inference:
Volume 1: Probability
by *J.G. Kalbfleisch*

Probability and Statistical Inference:
Volume 2: Statistical Inference
by *Nathan Keyfitz*

Graphical Exploratory Data Analysis
by *S. du Toit, et al.*

Counting for Something: An Historical
View of Statistics
by *William Peters*

S.H.C. du Toit   A.G.W. Steyn   R.H. Stumpf

Graphical Exploratory Data Analysis

With 180 Graphical Representations

Springer-Verlag New York Berlin Heidelberg
London Paris Tokyo

S.H.C. du Toit
Institute for Statistical Research
Human Sciences Research Council
Pretoria
South Africa

A.G.W. Steyn
Contract Researcher
Institute for Statistical Research
Human Sciences Research Council
Pretoria
South Africa

R.H. Stumpf
Contract Researcher
Institute for Statistical Research
Human Sciences Research Council
Pretoria
South Africa

AMS Classification: 62H99

Library of Congress Cataloging in Publication Data
du Toit, S.H.C.
    Graphical exploratory data analysis.
    (Springer texts in statistics)
    Bibliography: p.
    Includes index.
    1. Statistics—Graphic methods.   I. Steyn, A.G.W.
II. Stumpf, R.H.   III. Title.   IV. Series.
QA276.3.D778   1986   001.4′226   86-4009

Typeset by Asco Trade Typesetting, Ltd., Hong Kong.
Printed and bound by R.R. Donnelley and Sons, Harrisonburg, Virginia.
Printed in the United States of America.

9 8 7 6 5 4 3 2 1

ISBN 0-387-96313-8 Springer-Verlag New York Berlin Heidelberg
ISBN 3-540-96313-8 Springer-Verlag Berlin Heidelberg New York

Preface

Portraying data graphically certainly contributes toward a clearer and more penetrative understanding of data and also makes sophisticated statistical data analyses more marketable. This realization has emerged from many years of experience in teaching students, in research, and especially from engaging in statistical consulting work in a variety of subject fields.

Consequently, we were somewhat surprised to discover that a comprehensive, yet simple presentation of graphical exploratory techniques for the data analyst was not available. Generally books on the subject were either too incomplete, stopping at a histogram or pie chart, or were too technical and specialized and not linked to readily available computer programs. Many of these graphical techniques have furthermore only recently appeared in statistical journals and are thus not easily accessible to the statistically unsophisticated data analyst.

This book, therefore, attempts to give a sound overview of most of the well-known and widely used methods of analyzing and portraying data graphically. Throughout the book the emphasis is on exploratory techniques. Realizing the futility of presenting these methods without the necessary computer programs to actually perform them, we endeavored to provide working computer programs in almost every case. Graphic representations are illustrated throughout by making use of real-life data. Two such data sets are frequently used throughout the text. In realizing the aims set out above we avoided intricate theoretical derivations and explanations but we nevertheless are convinced that this book will be of inestimable value even to a trained statistician.

We certainly do not wish to claim that this book represents an exhaustive treatment of the topic of graphical exploratory techniques. Due to the many graphical techniques currently in existence and those having been developed recently, we were forced to be selective in our choice of techniques presented here. While acknowledging the existence of those techniques not included in this book, we nevertheless believe that our choice of techniques presents a good cross section which should be of great benefit to every data analyst whichever discipline he may represent.

The origin of this book can be traced back to a series of lectures on graphical techniques presented in 1983 to researchers at the Human Sciences Research Council (HSRC) in the Republic of South Africa. These lectures eventually led to a seminar on graphical techniques and finally to a HSRC report in 1984. The HSRC graciously provided some financial support so that this report in turn, after much editing, changing, and new research led to the writing of this book.

In an undertaking of this nature various people and organizations usually play an indispensable role. First, we would like to thank the HSRC and the University of Pretoria for placing their computers at our disposal. Second, we would like to thank Dr. Nico Crowther, Director of the Institute for Statistical Research at the HSRC, for his constant encouragement and advice in completing this project. Furthermore, Terry Shaw and Arien Strasheim were of great help in editing the many computer programs as were Antoinette van der Merwe and Jacques Pieterse in helping to photocopy, check data, proofread, and so forth.

Dr. Trevor Hastie from the Institute for Biostatistics at the Medical Research Council in Cape Town was nice enough to write the initial draft on scatterplot smoothing. We however accept full responsibility for the final version of this section. We would also like to thank Professor Stephen Fienberg, advisor to Springer-Verlag, who was of great help to us during his recent visit to South Africa. His mature insight and stimulating suggestions led to many improvements in the final form of the book.

Lynette Hearne and Mynie Stobbe did an excellent job in preparing the many figures while Trisia Badenhorst and Christa de Bruin typed this difficult manuscript.

Furthermore, we wish to express our gratitude to the Biometrika Trust for permission to publish Table 28 from Biometrika Tables for Statisticians (1976) and to Professor W.J. Serfontein and Mr. I.B. Ubbink from the University of Pretoria for supplying us with the fitness/cholesterol data. Many of the other data sets used in this book are done so with the kind permission of the HSRC.

Finally, and most important of all, we would like to thank our wives Dorothy, Jeanetta, and Adie. Without their understanding and support we would not have been able to complete this book.

STEPHEN DU TOIT
GERT STEYN
ROLF STUMPF

Contents

CHAPTER 1

The Role of Graphics in Data Exploration

1. Introduction

One of the most difficult tasks of a researcher is to convey findings based on statistical analyses to interested persons. Failure to communicate these findings successfully puts paid to all his data-analytical work, irrespective of its quality.

Statistical communication can be effected through words (spoken or written), tables or through graphics (diagrams, pictures, charts, illustrations or schematic representations). The efficient conveying of a statistical message should involve the use of all three communication media, which are described by Mahon (1977) respectively as the infantry, artillery and cavalry of the statistical defense force. Although the effectiveness of each of these methods depends on the contents of the message, they should be used to supplement one another. If words (including numerical analyses) convey the statistical realities in too complex a way, tables may perhaps do the job more effectively and concisely. Graphics, on the other hand, are essential for conveying relations and trends to the statistical layman in an informal and simplified visual form.

Graphics may also be regarded as a data-exploring technique through which the researcher familiarizes himself with the data, consequently enabling him to make conditional evaluations. Tukey (1977) regards graphic methods as data detection in which clues are sought which can be used in the court of inferential statistics, with a view to make certain justifiable pronouncements.

Although diagrams, etc. are often appropriate for imparting important facts contained in a jungle of figures, one must be cautious of misrepresentations of data which may leave the wrong impression, for instance when only certain aspects of a data set are presented in isolation. The following is a typical example of a misleading graphic representation. Wainer (1984) presents examples of a number of such misleading representations.

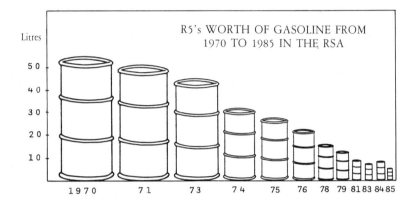

The following table shows the scores (out of 25) of four pupils in three so-called ability tests (Number Series Test at Educational Levels I, II and III. (For further particulars see Section 4.)

| Number Series Test | Education Level | | |
Pupil	I	II	III
1	19	15	19
2	18	11	15
3	21	16	19
4	21	8	13

Although the above table conveys the "facts" fairly clearly, namely that at Educational Level II the pupils obtained lower scores than at Educational Levels I and III and that Pupil 4 dropped out upon reaching Education Levels II and III, the figure below brings home this "message" even more effectively.

The effect and significance of graphics should therefore not be underestimated. The practice of consigning graphics to an appendix of a document, thereby implying that they are not essential for grasping the message, can not be recommended. Graphics, as a matter of fact, frequently manage to liven up and at the same time elucidate a highly technical and possibly dull message. The figure below also illustrates a common and advantageous characteristic of many graphical representations, namely that few if any calculations are required.

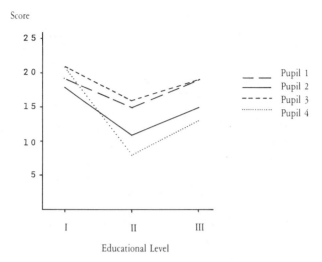

2. Historical Background

William Playfair, one of the founders of statistical graphics, wrote the following in 1801:

> For no study is less alluring or more dry and tedious than statistics, unless the mind and imagination are set to work or that the person studying is particularly interested in the subject; which is seldom the case with young men in any rank in life.

Playfair did much to make statistics more alluring, particularly through the propagation of graphic methods. Although certain graphic methods date back to the 10th and 11th century, it was mainly the work of Crome and Playfair (they, for instance, introduced bar diagrams and pie charts) toward the end of the 18th century that established graphics as an acceptable statistical method. Initially, statistical graphics were used to popularly present economic and population statistics. Well-known successors to Playfair in the development of statistical graphics include Bessel, Fourier (cumulative frequency polygon), and Quetelet (frequency polygon, histogram). The first statistical map was published by Fletcher in a statistical journal in 1849, whereas Florence Nightingale in 1857 used the so-called Coxcomb Chart to provide monthly data on casualties in the British army during the Crimean War. In 1874 Walker produced *The Statistical Atlas of the United States Based on the Results of the Ninth Census*. This publication was the first to contain examples of a variety of graphical displays.

The past 20 years have seen a considerable number of new initiatives in the field of graphical displays, particularly with regard to multivariate data.

Fienberg (1979) during a comprehensive investigation which involved study-ing the past 60 year's issues of two well-known statistical journals, *JASA* (*Journal of the American Statistical Association*) and *Biometrika*, made the alarming discovery however, that during this period there had been a system-atic reduction in the space devoted to graphical displays. This may have been due to the relative increase in theoretical inferential statistics during the twenties, which was pioneered by Fisher, Neyman, and Pearson. It would thus seem that there are indications of a gradual shift away from descriptive statistics in which statistical graphics played a major role, during the last number of decades.

Despite the above trend noticed by Fienberg, interest in statistical graphics has increased markedly during recent years, and fields of specialization within the general field of statistical graphics have already been the theme of several international conferences. As more computer programs are made available for the application of more sophisticated graphic techniques, exploratory graphical displays should, within the foreseeable future, develop into a major statistical field.

3. Content of the Book

Chapter 2 provides a brief review of basic descriptive graphics as applied to univariate and bivariate data. Many of the representations in this chapter are well-known and are included only for the sake of completeness. The stem-and-leaf plot as well as the box-and-whisker plot of Tukey are currently prominent in the field of descriptive statistics because they offer a natural, easy and interesting representation of a univariate data set. These two representations need wider introduction and are therefore discussed in greater detail.

Selecting an appropriate probability model to be used as the underlying structure for an observed data set is essential for the application of many inferential techniques. Graphical selection methods for some important uni-variate and multivariate probability models are dealt with in Chapter 3.

Various ingenious methods of representing multivariate data in two dimen-sions have been developed during the past two decades. The majority of these methods such as Chernoff faces and Andrews' curves ensure very vivid graphics. Chapter 4 gives a review of a number of these techniques.

A researcher often wants to divide the data points (cases) or the variables into fairly homogeneous groups. Cluster analysis is the obvious method for this and is discussed in Chapter 5. Attention is given particularly to hierarchical techniques and dendrogram representations but less well-known concepts such as digraphs and spanning trees are also introduced briefly in this chapter. Chapter 6 singles out certain matters relating to the broad subject of multi-dimensional scaling. Although the primary aim here, as in Chapter 4, is also to represent multidimensional data in a lower dimension, the results of more complex techniques in which interactions between the variables play an important role, are graphically demonstrated in this chapter.

Regression analysis is surely one of the most used techniques in statistics. Scatterplots and the representation of different residuals contribute significantly to the interpretation of regression results. Chapter 7 is largely devoted to this aspect, but also includes graphics in respect of techniques such as ridge regression. The CHAID and XAID procedures, which point out interactions between variables and which are particularly useful in exploratory studies involving large data sets, are discussed in Chapter 8. Chapter 9 deals with control charts. Although this type of representation is frequently used in quality control, control charts and in particular CUSUM charts are used more generally in other fields of application nowadays.

Chapter 10 is devoted to the graphical representation of time series in both the time and frequency domain. Finally, a number of other graphic techniques, divergent in nature, but nevertheless very important, are briefly discussed in Chapter 11.

4. Central Data Sets

The graphic techniques which are discussed, are illustrated as far as possible on the basis of actual data sets. Two data sets are frequently used and are briefly described next.

A. Ability Test Data

The first data set consists of data obtained during a long-term research project in South Africa called Project Talent Survey. This project was started in 1965 with the aim of determining the country's manpower potential and of making information available to enable this potential to be developed to a maximum. The test battery used in this project was compiled by the Institute for Psychometric and Edumetric Research of the Human Sciences Research Council in South Africa and consists of various ability tests. The data set consists of 21 variables measured in respect of approximately 2800 white pupils, with the first 18 variables constituting the scores (out of 25) obtained by the pupils in the various ability tests. The various test scores were obtained at three distinct educational levels. Educational Level I refers to 8 years of schooling, Educational Level II to 10 years of schooling and Educational Level III to 12 years of schooling.

The variables (tests) are:

X_1 = Number Series   Educational Level I
X_2 = Figure Analogies

Table 1.1. Subdata set of 28 pupils from the Ability Test Data set

Group	X_1	X_2	X_3	X_4	X_5	X_6	X_7	X_8	X_9	X_{10}	X_{11}	X_{12}	X_{13}	X_{14}	X_{15}	X_{16}	X_{17}	X_{18}	X_{19}	X_{20}	X_{21}
11	19	21	21	18	20	21	15	14	15	13	15	16	19	19	19	20	17	17	39	21	09
11	21	20	15	24	22	18	11	18	16	19	14	17	21	15	17	18	18	19	42	22	04
00	18	19	16	18	18	23	11	13	13	15	11	11	15	18	13	15	18	13	38	18	06
00	18	23	10	18	16	16	11	09	08	15	06	09	12	16	08	13	09	15	34	15	09
00	24	24	19	20	23	24	22	18	16	19	16	19	19	21	21	20	18	20	46	21	02
01	19	19	23	21	23	23	09	08	13	15	20	15	17	12	20	16	16	21	35	18	09
00	21	20	19	21	21	23	11	16	11	18	18	14	21	17	14	19	18	16	42	19	04
00	21	20	21	20	16	22	07	11	17	16	08	10	13	17	16	17	15	11	28	20	09
01	19	20	19	22	18	21	11	12	07	15	09	11	13	12	13	17	12	13	38	19	08
01	19	23	22	22	16	25	12	15	16	19	15	10	15	20	18	17	17	13	33	18	09
00	17	13	08	18	13	18	12	08	09	12	12	11	09	14	15	12	13	09	23	16	09
10	21	22	22	15	23	23	16	12	16	15	13	14	19	17	16	18	19	18	42	24	09
11	18	18	17	16	15	22	08	11	10	16	08	14	10	13	10	14	09	14	27	14	08
10	13	18	21	16	17	15	11	12	11	09	11	11	16	18	14	13	15	18	24	15	09
11	17	13	17	20	22	19	15	11	11	12	11	13	15	15	15	13	16	12	33	10	09
10	18	12	09	09	15	17	09	05	03	12	07	07	12	10	10	13	10	12	27	12	09
10	22	15	24	17	15	20	10	12	12	11	09	12	19	16	16	08	11	17	30	22	06
10	18	17	18	18	13	18	14	10	15	11	10	09	21	14	12	15	11	13	31	16	09
10	17	15	14	14	12	13	09	10	11	09	07	11	13	15	11	13	10	13	31	11	09
10	16	20	17	13	15	16	10	16	12	10	07	13	12	18	13	18	10	15	34	16	09
01	24	21	22	21	21	25	11	17	17	21	11	15	15	18	16	16	17	17	37	21	03
00	23	23	21	22	16	21	10	18	16	14	14	13	17	18	19	16	17	19	34	12	09
01	22	22	21	24	18	24	06	16	14	20	16	18	12	21	13	18	19	21	31	19	04
00	22	17	19	19	21	20	17	15	09	13	16	17	18	12	18	16	18	14	40	18	01
00	20	23	23	22	22	24	11	18	16	16	16	20	13	11	21	18	19	20	41	21	03
01	22	17	21	17	17	22	10	14	16	16	13	08	13	16	12	12	13	15	33	14	09
01	21	18	20	23	21	22	08	15	09	17	11	13	13	18	20	21	15	20	38	21	06
01	21	22	19	20	18	17	11	15	12	14	11	10	11	13	14	14	15	14	39	12	09

Groups: 00—Language Group A, boys; 01—Language Group A, girls; 10—Language Group B, boys; 11—Language Group B, girls.

X_3 = Pattern Completion
X_4 = Classification (Word Pairs)
X_5 = Verbal Reasoning
X_6 = Word Analogies

X_7 = Number Series   Education Level II
X_8 = Figure Analogies
X_9 = Pattern Completion
X_{10} = Classification (Word Pairs)
X_{11} = Verbal Reasoning
X_{12} = Word Analogies

X_{13} = Number Series   Educational Level III
X_{14} = Figure Analogies
X_{15} = Pattern Completion
X_{16} = Classification (Word Pairs)
X_{17} = Verbal Reasoning
X_{18} = Word Analogies

X_{19} = Junior Aptitude Test (Reasoning) (Out of 50)
X_{20} = Senior Aptitude Test (Verbal) (Out of 25)
X_{21} = Proposed Study Field

In addition to these variables the pupils are divided into four groups according to language and sex. The group in which a specific pupil falls can be regarded as a further variable.

The data for a random sample of 28 pupils from this data set appear in Table 1.1. This subset is frequently used in examples later.

B. Fitness/Cholesterol Data

Two of the many factors which are known to have some influence or relevance on the condition of the human heart are physical fitness and blood cholesterol level. In a related research project of the Department of Chemical Pathology of the University of Pretoria, four different homogeneous groups of adult males were considered. A large number of plasma lipid parameters were measured on each of the 66 individuals and fitness parameters were also measured on three of the four groups. The groups, coded by X_1, are:

$X_1 = 1$ Weightlifter ($n_1 = 17$)
    2 Student (control) ($n_2 = 20$)
    3 Marathon athlete ($n_3 = 20$)
    4 Coronary patient ($n_4 = 9$)

The characteristics that we will consider are:

Fitness characteristics

X_2 = Age (years)
X_3 = Length (cm)
X_4 = Mass (kg)
X_5 = Percentage fat
X_6 = Soma type 1
X_7 = Soma type 2
X_8 = Soma type 3
X_9 = Strength-breast (1b)
X_{10} = Strength-back (1b)
X_{11} = Reaction time—eye/right hand (seconds)
X_{12} = Reaction time—eye/left hand (seconds)
X_{13} = Reaction time—eye/right foot (seconds)
X_{14} = Reaction time—eye/left foot (seconds)
X_{15} = Strength (kg/sec)

Plasma lipid variable description

X_{16} = Triglycerides
X_{17} = Cholesterol (total)
X_{18} = HDL (ppt)
X_{19} = VLDL—cholesterol
X_{20} = LDL—cholesterol
X_{21} = HDL—cholesterol
X_{22} = HDL$_2$—cholesterol
X_{23} = HDL$_{3-v}$—cholesterol
X_{24} = HDL$_v$—cholesterol
X_{25} = HDL (ppt)/VLDL
X_{26} = HDL$_2$/HDL$_{3-v}$

The plasma lipid variables will be referred to in more popular terms as "cholesterol variables." The Fitness/Cholesterol Data are given in Table 1.2.

5. Different Types of Data

Data that have been collected can be quantitative (numerical) or qualitative (categorical). The former can be either discrete or continuous. Discrete data are observations which by nature can only assume fixed isolated values (size of family, years of training), whereas continuous data are observations which are inherently able to assume all possible values within a particular logical interval (family income, mass).

A considerable amount of information concerning persons or entities cannot be measured numerically but indicates only a quality. Observations

Table 1.2. Fitness/Cholesterol Data

OBS	X_1	X_2	X_3	X_4	X_5	X_6	X_7	X_8	X_9	X_{10}	X_{11}	X_{12}	Group 1 X_{13}	X_{14}	X_{15}	X_{16}	X_{17}	X_{18}	X_{19}	X_{20}	X_{21}	X_{22}	X_{23}	X_{24}	X_{25}	X_{26}
1	1	22	179.2	107.1	15.2	3.0	8.8	1.0	92	130	19	15	24	24	183.6	0.58	4.44	1.22	0.62	2.36	1.45	0.57	0.59	0.29	1.96	0.96
2	1	30	183.0	112.2	20.3	4.6	9.9	1.0	92	114	18	19	22	23	195.6	1.51	4.88	0.84	0.91	3.07	0.91	0.18	0.35	0.38	0.92	0.51
3	1	26	175.7	78.0	17.5	3.7	6.6	1.7	85	62	18	22	23	24	106.4	1.20	4.33	1.02	1.09	2.08	1.17	0.44	0.42	0.31	0.93	1.04
4	1	23	182.5	79.7	16.1	3.3	2.3	2.7	59	69	21	19	23	23	124.8	0.75	3.66	1.29	0.46	1.76	1.45	0.43	0.74	0.27	2.80	0.58
5	1	29	178.0	81.8	14.1	2.7	6.8	1.6	81	83	19	18	24	23	141.4	0.75	4.57	1.04	1.09	2.51	0.97	0.32	0.43	0.22	0.95	0.74
6	1	26	169.8	78.0	10.2	1.9	8.2	0.8	78	80	16	17	18	15	136.0	0.33	3.90	1.42	0.76	1.76	1.35	0.43	0.67	0.24	1.86	0.64
7	1	21	178.6	81.1	8.7	1.5	6.3	1.8	85	92	18	19	25	21	150.2	0.48	3.91	1.05	0.14	2.72	1.04	0.34	0.46	0.24	7.50	0.73
8	1	33	179.2	83.2	8.3	1.5	7.7	1.7	81	79	20	17	22	25	146.2	1.61	4.43	1.17	0.58	2.66	1.20	0.48	0.45	0.26	2.01	1.06
9	1	36	185.2	87.8	23.8	6.0	4.5	2.1	70	73	19	21	24	23	136.1	1.42	5.33	1.00	1.00	3.38	0.95	0.18	0.45	0.33	1.00	0.40
10	1	23	179.6	80.3	11.7	2.2	6.5	2.1	62	52	18	19	26	26	143.3	1.08	3.76	0.89	0.66	2.24	0.87	0.18	0.47	0.22	1.34	0.38
11	1	25	173.1	72.8	10.4	1.9	6.8	2.0	75	75	17	16	22	22	134.8	0.49	6.57	1.09	1.18	4.73	1.22	0.41	0.51	0.30	0.92	0.80
12	1	24	172.7	84.1	13.2	2.6	7.5	0.6	80	63	18	18	23	23	130.3	0.39	4.66	1.25	0.98	2.40	1.03	0.43	0.55	0.30	1.27	0.78
13	1	30	172.4	71.4	13.9	2.7	7.2	2.0	71	67	14	17	20	25	128.4	0.44	4.70	1.17	0.38	2.89	1.43	0.45	0.65	0.33	3.07	0.69
14	1	24	173.8	70.1	9.8	1.8	6.6	2.5	93	97	21	19	17	15	115.1	0.65	6.10	1.05	0.52	3.98	1.31	0.59	0.46	0.26	2.01	1.28
15	1	36	183.9	98.9	21.4	5.3	7.9	0.8	72	68	24	24	21	17	173.8	1.48	5.23	1.04	0.71	3.53	0.99	0.20	0.42	0.38	1.46	0.47
16	1		179.9	77.6	11.9	2.3	6.5	2.5	76	89	20	21	28	27	130.8	0.90	4.49	1.00	0.94	2.60	0.97	0.32	0.41	0.24	1.06	0.78
17	1	26	179.2	87.1	19.4	4.5	6.8	1.1	73	71	19	20	26	24	137.8	0.76	3.72	0.96	0.58	2.42	0.71	0.18	0.33	0.21	1.65	0.54

Table 1.2 (continued)

OBS	X_1	X_2	X_3	X_4	X_5	X_6	X_7	X_8	X_9	X_{10}	X_{11}	X_{12}	X_{13}	X_{14}	X_{15}	X_{16}	X_{17}	X_{18}	X_{19}	X_{20}	X_{21}	X_{22}	X_{23}	X_{24}	X_{25}	X_{26}
														Group 2												
18	2	28	186.1	80.4	18.2	3.9	4.4	3.2	73	54	17	16	23	24	125.9	0.78	5.05	0.85	0.73	3.59	0.74	0.18	0.32	0.24	1.16	0.56
19	2	21	184.4	69.8	11.3	2.1	4.7	4.4	115	77	18	14	26	25	124.6	0.97	2.97	1.13	0.46	1.78	0.74	0.22	0.35	0.17	2.45	0.62
20	2	22	179.2	74.7	18.6	4.1	5.6	2.8	70	66	19	18	25	26	125.9	0.46	2.87	1.05	0.20	1.71	0.97	0.32	0.44	0.21	5.25	0.72
21	2	22	179.9	76.5	18.2	4.1	5.3	2.6	.	.	16	17	21	22	130.1	0.70	4.27	1.01	0.88	2.33	1.07	0.27	0.51	0.29	1.14	0.52
22	2	35	192.6	95.0	14.5	2.9	5.1	2.5	65	70	15	20	23	24	158.8	0.70	3.71	1.06	0.52	2.10	1.12	0.35	0.48	0.29	2.03	0.72
23	2	26	172.2	64.8	17.3	3.8	5.2	3.0	65	53	21	20	23	27	107.3	0.47	4.56	0.97	1.03	2.60	0.94	0.20	0.37	0.37	0.94	0.54
24	2	23	172.7	72.8	16.1	3.3	6.5	1.9	66	55	18	19	22	29	125.5	0.85	4.56	1.11	0.80	2.90	0.85	0.30	0.42	0.13	1.38	0.71
25	2	24	180.7	80.8	14.1	2.8	6.3	2.2	113	110	19	16	24	25	136.2	0.89	3.80	1.25	0.59	1.82	1.41	0.67	0.48	0.26	2.11	1.39
26	2	30	177.1	81.6	17.4	3.7	6.5	1.5	70	65	19	19	27	25	139.9	1.15	5.30	1.13	0.61	3.37	1.33	0.46	0.62	0.25	1.85	0.74
27	2	22	168.2	62.4	14.6	2.8	5.2	2.7	34	60	16	15	21	25	109.6	0.97	4.56	0.97	0.73	2.78	1.05	0.37	0.45	0.23	1.32	0.82
28	2	23	180.1	78.2	20.4	4.7	4.8	2.5	53	56	18	19	26	27	124.8	0.90	5.24	0.81	0.45	4.06	0.76	0.13	0.37	0.26	1.80	0.35
29	2	21	174.6	69.2	12.8	2.4	5.4	2.8	52	55	16	17	21	17	116.7	0.72	3.78	0.93	0.32	2.50	0.97	0.23	0.34	0.40	2.90	0.67
30	2	27	181.5	61.0	12.0	2.3	3.0	5.4	44	40	19	17	28	26	92.6	1.08	4.18	0.84	0.57	2.73	0.89	0.35	0.26	0.29	1.47	1.34
31	2	30	184.8	70.0	7.6	1.4	3.7	4.4	67	43	20	19	27	24	107.4	0.78	4.75	1.00	0.47	2.76	1.52	0.69	0.45	0.38	2.12	1.53
32	2	21	189.5	73.9	19.4	4.4	2.9	4.7	55	57	21	19	25	26	113.3	0.74	3.01	1.01	0.57	1.57	0.87	0.17	0.51	0.19	1.77	0.33
33	2	24	174.3	73.7	20.4	4.7	5.7	2.1	77	86	19	18	23	24	119.9	2.72	5.81	0.80	0.65	4.33	0.83	0.11	0.44	0.28	1.23	0.25
34	2	27	183.5	60.7	10.1	1.8	5.8	2.9	47	37	19	18	25	27	96.9	0.92	3.61	1.30	0.40	1.71	1.50	0.50	0.77	0.23	3.25	0.64
35	2	29	173.0	65.9	18.1	3.8	5.2	2.9	70	54	16	21	25	25	101.1	0.82	5.94	1.29	0.35	4.06	1.53	0.36	0.85	0.32	3.68	0.42
36	2	33	178.1	71.1	16.3	3.3	5.5	3.1	74	54	19	19	26	26	112.4	1.75	6.14	0.97	0.66	4.50	0.98	0.16	0.44	0.38	1.46	0.36
37	2	33	182.7	78.9	14.9	2.9	4.9	2.8	87	63	16	17	25	24	133.0	2.66	6.04	0.69	1.29	3.98	0.77	0.36	0.07	0.34	0.53	5.14

Group 3

ID	Grp																									
38	3	45	182.4	89.5	21.7	5.3	6.6	1.5	60	65	18	18	23	21	122.1	0.86	5.47	1.16	0.71	3.01	1.76	0.62	0.84	0.31	1.63	0.73
39	3	23	176.4	58.8	10.5	1.9	4.0	4.8	51	42	26	21	30	33	93.0	0.84	4.66	0.94	0.51	2.93	1.23	0.31	0.58	0.35	1.84	0.53
40	3	27	169.1	70.7	24.0	5.8	4.8	1.6	49	41	19	23	·	32	·	1.15	7.19	1.15	1.48	4.47	1.22	0.26	0.53	0.44	0.77	0.49
41	3	37	173.7	76.4	16.6	3.6	6.7	1.6	67	78	18	20	23	24	112.2	1.12	4.88	1.20	1.02	2.43	1.42	0.55	0.61	0.27	1.17	0.90
42	3	18	190.7	70.7	13.1	2.7	3.5	5.4	54	40	17	20	25	24	108.4	0.72	4.42	1.29	0.28	2.60	1.57	0.64	0.64	0.29	4.60	1.00
43	3	29	185.9	71.0	15.8	3.2	3.5	4.5	45	42	18	18	25	28	106.6	1.62	4.70	0.94	0.47	2.64	1.59	0.39	0.62	0.59	2.00	0.62
44	3	43	177.8	69.4	13.9	2.8	4.8	3.3	55	52	21	23	28	28	92.2	1.23	6.12	1.11	0.71	4.33	1.08	0.21	0.47	0.40	1.56	0.44
45	3	33	171.6	67.5	14.0	2.8	4.6	2.5	54	55	17	17	28	22	110.8	1.22	6.17	1.37	0.61	4.02	1.53	0.46	0.63	0.40	2.24	0.73
46	3	30	184.5	74.2	13.2	2.6	3.9	3.8	67	67	21	18	22	22	116.1	1.13	5.10	1.13	0.65	2.94	1.51	0.44	0.79	0.39	1.73	0.55
47	3	23	174.0	69.0	14.6	2.9	4.4	2.7	68	70	19	21	27	22	109.1	0.98	5.10	1.11	0.59	3.42	1.09	0.25	0.49	0.35	1.88	0.51
48	3	25	169.4	59.1	11.7	2.3	3.9	3.5	62	45	18	19	26	26	84.8	0.62	4.00	1.20	0.69	2.17	1.15	0.79	0.25	0.12	1.73	3.16
49	3	39	182.8	71.6	13.8	2.8	5.1	3.7	60	40	·	19	24	23	106.3	0.38	5.38	1.24	0.61	3.33	1.44	0.25	0.74	0.45	2.03	0.33
50	3	29	182.8	75.6	15.8	3.2	4.3	3.3	68	54	19	·	25	24	121.8	0.86	4.18	0.98	0.72	2.69	0.77	0.17	0.51	0.09	1.36	0.33
51	3	27	182.2	78.4	13.6	2.6	6.2	2.8	70	66	24	23	30	27	108.4	1.25	5.00	1.33	1.06	2.53	1.41	0.39	0.66	0.36	1.25	0.59
52	3	32	177.5	66.0	14.8	3.0	3.4	3.8	55	42	17	18	23	23	91.2	0.90	4.67	1.16	0.85	2.71	1.11	0.40	0.44	0.28	1.36	0.90
53	3	45	186.4	81.5	15.7	3.3	4.8	3.1	71	57	13	17	26	26	111.2	0.56	3.51	1.33	0.64	1.69	1.23	0.72	0.35	0.16	2.07	2.05
54	3	26	178.7	64.0	11.6	2.2	3.1	4.3	70	54	19	21	24	24	83.8	0.66	4.44	0.94	0.74	2.81	0.90	0.25	0.38	0.28	1.27	0.65
55	3	27	174.9	74.1	17.4	3.6	5.3	2.1	68	46	18	18	24	24	101.1	0.73	3.56	1.11	0.38	1.84	1.17	0.43	0.40	0.34	2.92	1.07
56	3	23	171.2	69.8	14.3	2.9	6.3	2.1	80	61	19	21	27	29	104.8	0.73	3.15	·	0.44	1.92	0.79	0.35	0.29	0.15	·	·
57	3	24	176.3	72.4	15.3	3.1	5.5	2.6	73	81	18	20	26	28	113.3	0.50	4.97	1.03	0.75	3.24	0.98	·	·	·	1.37	1.20

Group 4

ID	Grp																									
58	4	46	·	·	·	·	·	·	·	·	·	·	·	·	·	3.80	6.98	0.83	1.75	4.38	0.86	0.13	0.38	0.35	0.47	0.32
59	4	60	·	·	·	·	·	·	·	·	·	·	·	·	·	2.70	5.50	·	1.08	3.66	0.76	0.18	0.35	0.23	·	0.51
60	4	62	·	·	·	·	·	·	·	·	·	·	·	·	·	1.60	6.61	0.93	0.58	4.78	1.25	0.23	0.66	0.36	1.60	0.34
61	4	67	·	·	·	·	·	·	·	·	·	·	·	·	·	1.60	4.70	0.73	0.69	3.18	0.84	0.17	0.32	0.35	1.05	0.53
62	4	33	·	·	·	·	·	·	·	·	·	·	·	·	·	1.90	4.53	·	0.74	2.79	1.01	0.15	0.41	0.45	·	0.36
63	4	53	·	·	·	·	·	·	·	·	·	·	·	·	·	1.30	7.90	0.77	0.74	6.42	0.75	0.54	0.11	0.10	1.04	4.90
64	4	67	·	·	·	·	·	·	·	·	·	·	·	·	·	1.00	3.84	1.09	0.67	2.10	1.07	0.21	0.51	0.35	1.62	0.41
65	4	54	·	·	·	·	·	·	·	·	·	·	·	·	·	2.20	5.52	0.60	0.81	4.09	0.62	0.09	0.26	0.27	0.74	0.34
66	4	50	·	·	·	·	·	·	·	·	·	·	·	·	·	0.60	3.73	0.77	0.39	2.55	0.79	0.38	0.21	0.20	1.97	1.80

regarding a person's sex or language group are examples of qualitative or categorical data, which may be more definitely typified as nominal data as opposed to ordinal data, where the categories have a certain order. An example of the latter is data on a person's attitude toward a matter in question which may be classified as "strongly opposed," "opposed," "neutral," "in favor," "strongly in favor."

With regard to the Ability Test Data discussed in Section 4, it is clear that X_1 to X_{20} may be regarded as quantitative variables and X_{21} as well as "group" as categorical variables. Although X_1 to X_{20} all represent test results reproduced as integer values between 0 and 25 (0 and 50 for X_{19}) and are therefore strictly speaking discrete by nature, achievement measurements such as these are usually regarded as continuous.

6. Computer Programs

The analysis of multivariate data and producing the related graphics is often complex and usually cannot be tackled simply with a pocket calculator, a pencil, and graph paper. In this book the focus is mainly on procedures for which computer software is readily available. Although different computer program packages are used by way of illustration and an attempt is made each time to indicate both the input and the output, the aim is not to discuss the programs in detail since complete manuals are available for the program packages. Packages such as the SAS, BMDP and SPSS program series are indispensable nowadays for analyzing comprehensive data sets statistically.

CHAPTER 2

Graphics for Univariate and Bivariate Data

1. Introduction

This chapter deals with a number of diverse representations which are all aimed at providing meaningful, simple and interesting presentations of univariate and bivariate data. All these representations may be regarded as types of exploratory techniques that are used without any preconceived ideas of possible further analyses.

Section 2 is concerned with the basic and familiar techniques (such as the histogram) that are used to represent univariate data. Section 3 deals with the less common stem-and-leaf and box-and-whisker methods of representing univariate data. The representation of bivariate data is discussed in Section 4. Finally, a very important discussion regarding graphical perception is included in Section 5.

Figures 2.1–2.7 in Section 2, as well as Figure 2.9 in Section 4, were obtained through the SAS/GRAPH computer package. This package produces graphics on a color terminal screen, on the matrix printer with graph facilities or on a graph plotter.

The SAS/GRAPH procedures GCHART and GPLOT correspond to the SAS procedures CHART and PLOT, with all statements and options in the latter also available for use in GCHART and GPLOT. In addition to this, further options are available which enable SAS/GRAPH to produce more colorful and attractive visual representations. This system also has extensive titling and labelling facilities through which the graphic representations can be elucidated. For further information the reader is referred to the SAS/GRAPH 1985 User's Manual of the SAS Institute.

Table 2.1. Frequency distribution of the scores of 28 pupils at Educational Level II in the Number Series Test (X_7)

Scores		Frequency	Cumulative frequency
[5; 9)	////	4	4
[9; 13)	//// //// //// ///	18	22
[13; 17)	////	4	26
[17; 21)	/	1	27
[21; 25)	/	1	28
Total		28	

2. Graphics for Univariate Data

Consider the 28 observations of variable X_7 as shown in Table 1.1. The smallest and greatest values are, respectively, 9 and 22. A frequency distribution of the data is given in Table 2.1. The notation [5; 9) in effect indicates the interval 5–8, but is preferred to the latter since it stresses the continuity inherent in the data. The histogram of this frequency distribution is represented in Figure 2.1.

Input: SAS/GRAPH PROC GPLOT program for compilation of histogram

```
DATA ;
INPUT SCORE FREQ @@ ;
CARDS ;
1 4 2 18 3 4 4 1 5 1
;
PROC    GCHART ;
VBAR SCORE / SUBGROUP=SCORE SUMVAR=FREQ NOSYMBOL NOLEGEND VREF=0
             MIDPOINTS=1 2 3 4 5 CTEXT=GREEN CAXIS=ORANGE ;
TITLE ;
FOOTNOTE1 .F=DUPLEX .H=1 .J=L .C=VIOLET 'Figure 2.1 - Histogram of the scores of
 28 pupils ' ;
FOOTNOTE2 .F=DUPLEX .H=1 .J=L .C=VIOLET 'at Educational Level II in the Number S
eries Test ( X' .M=(-0.25,-0.25) .H=0.5 '7'.M=(+0,+0.25) .H=1 ' )' ;
```

Output:

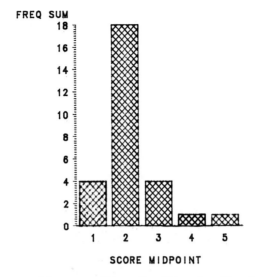

Figure 2.1. Histogram of the scores of 28 pupils at Educational Level II in the Number Series Test (X_7).

Figure 2.2 shows the frequency polygon of the frequency distribution given in Table 2.1 and can be regarded as a more continuous representation of the data set. Both the histogram and the frequency polygon show a distribution of scores which is concentrated in the interval [9; 13) but which reflects a degree of positive skewness.

Input: SAS/GRAPH PROC GPLOT program for compilation of frequency polygon

```
DATA ;
INPUT   SCORE NUMBER @@ ;
CARDS ;
0 0 3 0 7 4 11 18 15 4 19 1 23 1 27 0
;
PROC · GPLOT;
PLOT    NUMBER*SCORE/ NOLEGEND HAXIS=0 3 7 11 15 19 23 27 VREF=0 ;
SYMBOL1    I=JOIN V=NONE;
TITLE ;
FOOTNOTE1 .F=DUPLEX .H=1 .J=L 'Figure 2.2 - Frequency polygon of the scores of 2
8 pupils at' ;
FOOTNOTE2 .F=DUPLEX .H=1 .J=L 'Educational Level II in the Number Series Test (
X' .M=(-0.25,-0.25) .H=0.5 '7' .M=(+0,+0.25) .H=1 ')' ;
```

Output:

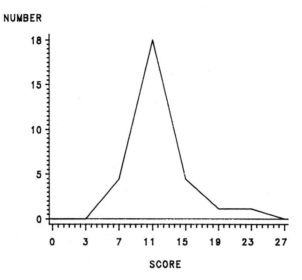

Figure 2.2. Frequency polygon of the scores of 28 pupils at Educational Level II in the Number Series Test (X_7).

The cumulative frequency polygon shown in Figure 2.3 can be used to estimate the number (or percentage) of pupils who obtained a score that is equal to or below a prescribed score. In this regard the median is usually vitally important and can be determined by reading the test score that corresponds to a cumulative frequency of 14 (i.e., 50 percent of 28) on the vertical axis.

Input: SAS/GRAPH PROC GPLOT program for compilation of cumulative frequency polygon

```
DATA ;
INPUT   SCORE NUMBER @@ ;
CARDS ;
0 0 5 0 9 4 13 22 17 26 21 27 25 28
;
PROC    GPLOT ;
PLOT    NUMBER*SCORE / NOLEGEND HAXIS=5 10 15 20 25 VREF=0 ;
SYMBOL1    I=JOIN V=NONE C=BLACK;
TITLE ;
FOOTNOTE1 .F=DUPLEX .H=1 .J=L 'Figure 2.3 - Cumulative frequency polygon of the
scores of 28' ;
FOOTNOTE2 .F=DUPLEX .H=1 .J=L 'pupils at Educational Level II in the Number Seri
es Test ( X' .M=(-0.25,-0.25) .H=0.5 '7' .M=(+0,+0.25) .H=1 ')' ;
```

Output:

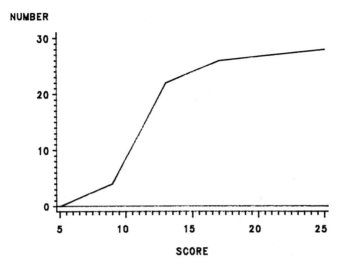

Figure 2.3. Cumulative frequency polygon of the scores of 28 pupils at Educational Level II in the Number Series Test (X_7).

Table 2.2. Arithmetic means of scores in Number Series Test (X_7)

Group	Frequency	I	Educational level II	III
Boys	16	19.31	11.94	15.31
Girls	12	20.17	10.58	14.67
Language group A	17	20.65	10.59	14.35
Language group B	11	18.18	11.64	16.09

Consider the achievements of the 28 pupils in three tests; namely X_1, X_7, and X_{13} (Number Series Test at Educational Levels I, II and III) as indicated in Table 1.1. If a distinction is drawn on the basis of language and sex and the arithmetic means of the scores for each group are calculated, the end result is the data shown in Table 2.2. The frequencies on which the means are based are also given in the table.

Figures 2.4 and 2.5 represent bar charts of the arithmetic means which have been constructed with a view to simultaneously pointing out differences in achievement between educational level and sex and between educational level and language group, respectively.

Input: SAS/GRAPH PROC GPLOT program for compilation of bar chart (According to educational level and sex of pupils)

```
PROC    FORMAT;
        VALUE S 1 = 'MALE'
                2 = 'FEM' ;
        VALUE L 1 = 'I'
                2 = 'II'
                3 = 'III' ;
DATA ;
LABEL LEVEL = 'LEVEL' ;
LABEL SEX   = 'SEX  ' ;
LABEL SCORE = 'SCORE' ;
INPUT LEVEL SEX SCORE ;
FORMAT SEX S. ;
FORMAT LEVEL L. ;
CARDS ;
1 1 19.31
1 2  0.0
1 1  0.0
1 2 20.17
2 1 11.94
2 2  0.0
2 1  0.0
2 2 10.58
3 1 15.31
3 2  0.0
3 1  0.0
3 2 14.67
;
PROC    GCHART ;
VBAR    SEX / SUMVAR=SCORE GROUP=LEVEL GSPACE=5 SPACE=0 REF=0 CAXIS=ORANGE
              CTEXT=GREEN MIDPOINTS=1 2 SUBGROUP=SEX ;
PATTERN1 V=R2 C=BLUE ;
PATTERN2 V=X2 C=RED ;
TITLE ;
FOOTNOTE1 .F=DUPLEX .H=1 .J=L .C=VIOLET 'Figure 2.4 (a) - Bar chart of arithmeti
c means' ;
FOOTNOTE2 .F=DUPLEX .H=1 .J=L .C=VIOLET 'of scores in the Number Series Test ( X
' .M=(-0.25,-0.25) .H=0.5 '7' .M=(+0.+0.25) .H=1 ')' ;
```

Output:

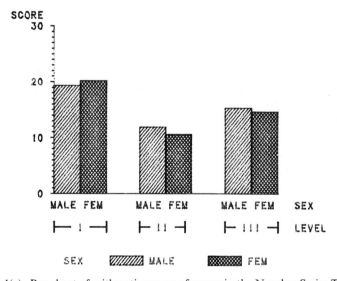

Figure 2.4(a). Bar chart of arithmetic means of scores in the Number Series Test (X_7).

Input: SAS/GRAPH PROC GCHART program for compilation of three-dimensional bar chart (according to educational level and sex of pupils)

```
PROC    FORMAT;
        VALUE G 1='MALE'
                2='FEMALE' ;
        VALUE S 6='LEVEL I'
                8='LEVEL II'
                10='LEVEL III' ;
DATA ;
INPUT   SEX STANDARD SCORE @@ ;
FORMAT SEX G. ;
FORMAT STANDARD S. ;
CARDS ;
1 6 19.31 2 6 20.17 1 8 11.94 2 8 10.58 1 10 15.31 2 10 14.67
;
PROC    GCHART ;
BLOCK   STANDARD / SUBGROUP=SEX GROUP=SEX SUMVAR=SCORE
                   MIDPOINTS=6 8 10 NOLEGEND ;
PATTERN1 C=BLACK V=L2 ;
PATTERN2 C=BLACK V=X2 ;
TITLE ;
FOOTNOTE1 .F=DUPLEX .H=1 .J=L 'Figure 2.4 (D) - Bar chart of arithmetic means of
  scores attained ' ;
FOOTNOTE2 .F=DUPLEX .H=1 .J=L 'by boys and girls in the Number Series Test ( X'
.M=(-0.25,-0.25) .H=0.5 '7' .M=(+0,+0.25) .H=1 ')' ;
```

Output:

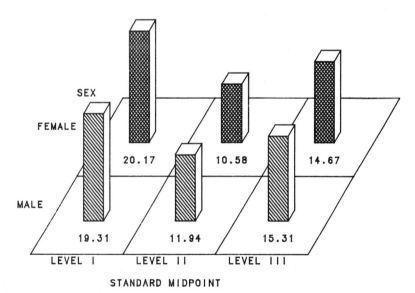

Figure 2.4(b). Bar chart of arithmetic means of scores attained by boys and girls in the Number Series Test (X_7).

If the SAS PROC CHART procedure is used instead to obtain a bar chart like the one given in Figure 2.4(b), the input is similar to the above input, the only difference being that the statement "PROC GCHART" is replaced by the statement "PROC CHART." In this case the output would be as in Figure 2.4(c). The same applies to the SAS/GRAPH procedure PROC GPLOT and the SAS procedure PROC PLOT.

Output: Bar chart (in three dimensions) from a SAS PROC CHART program

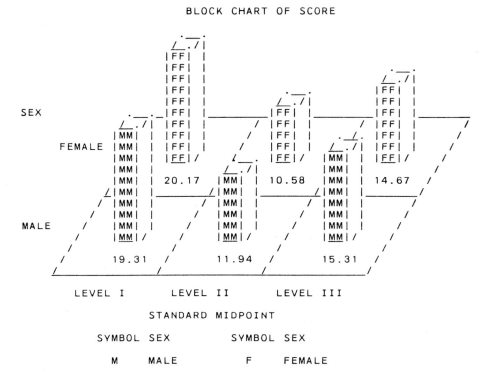

Figure 2.4(c). Bar chart of arithmetic means of scores in the Number Series Test (X_7).

Input: SAS/GRAPH PROC GPLOT program for compilation of bar chart (according to educational level and language group of pupils)

```
PROC    FORMAT;
        VALUE LA 1 = 'A'
                 2 = 'B' ;
        VALUE LE 1 = 'I'
                 2 = 'II'
                 3 = 'III' ;
DATA ;
LABEL LEVEL  = 'LEVEL' ;
LABEL LANG   = 'LANG ' ;
LABEL SCORE  = 'SCORE' ;
INPUT LEVEL LANG SCORE ;
FORMAT LANG  LA. ;
FORMAT LEVEL LE. ;
CARDS ;
1 1 20.62
1 2  0.00
1 1  0.00
1 2 18.18
2 1 10.59
2 2  0.00
2 1  0.00
2 2 11.64
3 1 14.35
3 2  0.00
3 1  0.00
3 2 16.09
;
PROC    GCHART ;
VBAR    LANG / SUMVAR=SCORE GROUP=LEVEL GSPACE=5 SPACE=0 MIDPOINTS=1 2
               SUBGROUP=LANG ;
PATTERN1 V=R2 C=BLACK ;
PATTERN2 V=X2 C=BLACK ;
TITLE ;
FOOTNOTE1 .F=DUPLEX .H=1 .J=L 'Figure 2.5 - Bar chart of arithmetic means of sco
res attained by' ;
FOOTNOTE2 .F=DUPLEX .H=1 .J=L 'pupils from Language Groups A and B in the Number
 Series Test  ( X' .M=(-0.5,-0.5) .H=0.5 '7' .M=(+0,+0.5) .H=1 ')' ;
```

Output:

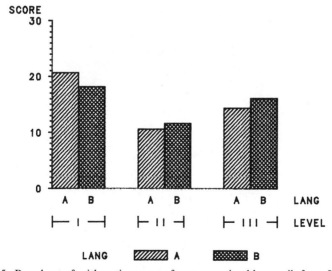

Figure 2.5. Bar chart of arithmetic means of scores attained by pupils from Language Groups A and B in the Number Series Test (X_7).

Figure 2.6 and 2.7 show percentage representations which, in a meaningful way, represent the composition of the 28 pupils according to language group and sex. These figures respectively show pie chart representations and a divided bar chart representation.

Input: SAS/GRAPH PROC GCHART program for compilation of pie chart (according to language group of pupils)

```
PROC    FORMAT ;
        VALUE L 1='LANG A'
                2='LANG B' ;
DATA ;
INPUT  LANG SCORE @@ ;
FORMAT LANG L. ;
FORMAT SCORE 4.1 ;
CARDS ;
1 60.7 2 39.3
;
PROC    GCHART ;
PIE     LANG / SUMVAR=SCORE MIDPOINTS=1 2 FILL=X ;
TITLE ;
FOOTNOTE1 .F=DUPLEX .H=1 .J=L 'Figure 2.6 (a) - Pie chart of the composition of
28 pupils' ;
FOOTNOTE2 .F=DUPLEX .H=1 .J=L ' with regard to language' ;
```

Output:

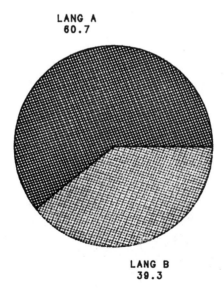

Figure 2.6(a). Pie chart of the composition of 28 pupils with regard to language.

Input: SAS/GRAPH PROC GCHART program for compilation of pie chart (according to sex and language group of pupils)

```
PROC    FORMAT;
        VALUE L 1='LANG A    MALE'
                2='LANG A FEMALE'
                3='LANG B    MALE'
                4='LANG B FEMALE' ;
DATA ;
INPUT   LANG SCORE @@ ;
FORMAT LANG L. ;
FORMAT SCORE 4.1 ;
CARDS ;
1 32.1 2 28.6 3 25.0 4 14.3
;
PROC    GCHART;
PIE     LANG / SUMVAR=SCORE FILL=X MIDPOINTS= 1 2 3 4 CTEXT=ORANGE ;
TITLE ;
FOOTNOTE1 .F=DUPLEX .H=1 .C=VIOLET .J=L 'Figure 2.6 (b) - Pie chart of the compo
sition of 28 pupils' ;
FOOTNOTE2 .F=DUPLEX .H=1 .C=VIOLET .J=L 'with regard to sex and language' ;
```

Output:

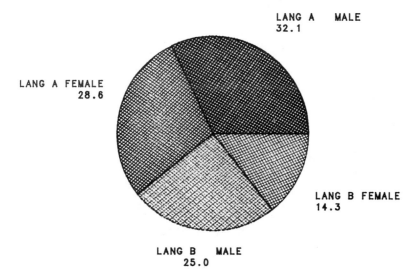

Figure 2.6(b). Pie chart of the composition of 28 pupils with regard to sex and language.

Input: SAS/GRAPH PROC GPLOT program for compilation of divided bar chart
(according to sex and language group of pupils)

```
PROC    FORMAT ;
        VALUE B 1=' L '
                2=' S '
                3='L/S' ;
DATA    DATA ;
INFILE  CARDS ;
INPUT   BAR PERC GROUP ;
CARDS ;
        1 60.7    1
        1 39.3    2
        1  0.0    3
        1  0.0    4
        2  0.0    1
        2  0.0    2
        2 57.1    3
        2 42.9    4
        3 32.1    1
        3 28.6    2
        3 14.3    3
        3 25.0    4
DATA    ANNO ;
        LENGTH X Y 8 XSYS YSYS $ 1 WHEN  $ 1 FUNCTION $ 8
        TEXT $ 8 POSITION $ 1  ;
        YSYS = '5' ;  XSYS = '5' ;
        FUNCTION = 'LABEL' ;
        WHEN = 'A' ;
        POSITION = '3' ;
        X = 40 ;      Y = 90 ;    TEXT = 'LANG A' ;   OUTPUT ;
        X = 40 ;      Y = 85 ;    TEXT = '39.3  ' ;   OUTPUT ;
        X = 40 ;      Y = 50 ;    TEXT = 'LANG B' ;   OUTPUT ;
        X = 40 ;      Y = 45 ;    TEXT = '60.7  ' ;   OUTPUT ;
        X = 60 ;      Y = 90 ;    TEXT = 'FEMALE' ;   OUTPUT ;
        X = 60 ;      Y = 85 ;    TEXT = '42.9  ' ;   OUTPUT ;
        X = 60 ;      Y = 50 ;    TEXT = 'MALE  ' ;   OUTPUT ;
        X = 60 ;      Y = 45 ;    TEXT = '57.1  ' ;   OUTPUT ;
        X = 80 ;      Y = 90 ;    TEXT = 'LANG B' ;   OUTPUT ;
        X = 80 ;      Y = 85 ;    TEXT = 'FEMALE' ;   OUTPUT ;
        X = 80 ;      Y = 80 ;    TEXT = '25.0  ' ;   OUTPUT ;
        X = 80 ;      Y = 70 ;    TEXT = 'LANG B' ;   OUTPUT ;
        X = 80 ;      Y = 65 ;    TEXT = 'MALE  ' ;   OUTPUT ;
        X = 80 ;      Y = 60 ;    TEXT = '14.3  ' ;   OUTPUT ;
        X = 80 ;      Y = 50 ;    TEXT = 'LANG A' ;   OUTPUT ;
        X = 80 ;      Y = 45 ;    TEXT = 'FEMALE' ;   OUTPUT ;
        X = 80 ;      Y = 40 ;    TEXT = '28.6  ' ;   OUTPUT ;
        X = 80 ;      Y = 30 ;    TEXT = 'LANG A' ;   OUTPUT ;
        X = 80 ;      Y = 25 ;    TEXT = 'MALE  ' ;   OUTPUT ;
        X = 80 ;      Y = 20 ;    TEXT = '32.1  ' ;   OUTPUT ;
PROC    GCHART DATA=DATA  ANNOTATE=ANNO ;
VBAR    BAR / SPACE=10 SUBGROUP=GROUP
              SUMVAR=PERC DISCRETE NOLEGEND ;
FORMAT BAR B. ;
PATTERN1 V=L1 C=BLACK ;
PATTERN2 V=X1 C=BLACK ;
PATTERN3 V=L1 C=BLACK ;
PATTERN4 V=X1 C=BLACK ;
TITLE ;
FOOTNOTE1 .F=DUPLEX .H=2 .J=L 'Figure 2.7 - Bar chart of the composition of' ;
FOOTNOTE2 .F=DUPLEX .H=2 .J=L '28 pupils with regard to sex and language' ;
```

Output:

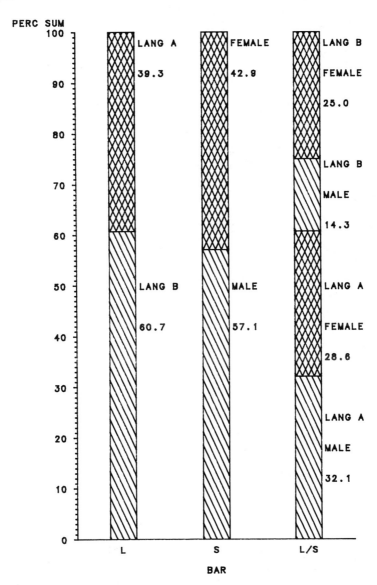

Figure 2.7. Bar chart of the composition of 28 pupils with regard to sex and language.

Comments

(a) Sometimes bar charts are compiled with a view to comparing ratios in populations of unequal size. This information can be incorporated in the representation by selecting the width of each bar proportional to the particular population size.

(b) Situations arise in which the percentages in the different categories of a variable vary over a wide range and whereby some percentages are smaller than one percent. In this case the percentages can be represented as triangles whereby a quadratic percentage scale is used.

3. Stem-and-Leaf Plots

In an investigation into the possible effect of television on school-going children, the intelligence coefficients of the children were obtained. Table 2.3 shows the coefficients of 70 Educational Level I boys, which vary between 91 and 137.

Table 2.4 shows a so-called stem-and-leaf plot of the above data. The figures to the left of the vertical line show the stems and those to the right are the leaves. The first IQ value of 103 is written to the right of 10∗ as 3, the second, namely 115 to the right of 11∗ as 5, etc. After all 70 values have been represented in this way and a tally has been made of the number of observations at each stem, the result is in fact a frequency distribution. The advantage

Table 2.3. IQs of 70 Educational Level I boys

103	120	124	109	103	107	107
115	103	114	119	115	119	103
124	117	120	105	110	110	96
137	121	105	96	112	116	110
98	123	91	97	111	127	132
115	132	97	119	96	112	103
94	114	115	109	110	98	120
110	119	122	115	99	122	105
99	128	117	127	116	102	103
117	121	127	117	110	100	103

Table 2.4. Stem-and-leaf plot of the IQs of 70 Educational Level I boys

		#
9∗	84917676986	11
10∗	3359593720733533	16
11∗	55077494579957502106090620	26
12∗	40138140277720	14
13∗	722	3
		70

Table 2.5. Ordered stem-and-leaf plot of the IQs of 70 Educational Level I boys

		#
9*	14666778899	11
10*	0233333335557799	16
11*	00000012244555556677779999	26
12*	00011223447778	14
13*	227	3
		70

of this distribution, in contrast to the distribution given in Table 2.1, is that the original information has been retained. If Table 2.4 is turned on its side we also have a histogram representation of the data. Table 2.5 again shows the stem-and-leaf plot but now the "leaves" of each "stem" have been ordered from small to large.

Table 2.5 can now be used to enumerate the quantiles. Since the values of all the observed data are available in Table 2.5 (but ordered from small to large), $i(70 + 1)/100$ is used to obtain the position of the ith percentile. This positional value can be determined through enumeration from the smallest observation, namely 91.

Consider for example:

$$\text{median} = \text{Value of observation in } \frac{50(70 + 1)}{100}\text{-th position}$$

$$= \text{Value of 35.5-th observation}$$
$$= 112$$

$q_1 \quad = \text{Value of 17.75-th observation}$
$\qquad = 103$

$q_3 \quad = \text{Value of 53.25-th observation}$
$\qquad = 119.25$

The stem-and-leaf plot can be arbitrarily adjusted to adapt to the unique nature of each data set. Consider the scores of the 28 Educational Level II pupils in the Number Series Test (X_7 in Table 1.1). After the analogy of Table 2.4 it would be possible to obtain a stem-and-leaf plot of the data as in Table 2.6(a). Tables 2.6(b) and (c) however, are more effective since they reduce the length of the intervals (or increase the number of stems). In Table 2.6(b) "*" indicates a 5, 6, 7, 8 or 9 and a "·" indicates 0, 1, 2, 3 or 4. In Table 2.6(c) "*" indicates a 0 or 1, "+" a 2 or 3, etc. The representations in Tables 2.6(b) and (c) are called stretched stem-and-leaf plots.

Both the SAS PROC UNIVARIATE and BMDP-2D programs can be used to obtain a stem-and-leaf plot. As an illustration, the BMDP-2D program was used to obtain a stretched stem-and-leaf plot of the data in Table 2.3.

Table 2.6. Stem-and-leaf plots of
the scores of 28 Educational
Level II pupils in the Number
Series Test

(a)			#
	*	9789968	7
1	*	51111122615040107101	20
2	*	2	1
			28

(b)			#
	*	9789968	7
1	.	1111122104010101	16
	*	5657	4
2	.	2	1
			28

(c)			#
	−	76	2
	.	98998	5
1	*	1111110010101	13
	+	22	2
	×	554	3
	−	67	2
	.		0
2	*		0
	+	2	1
			28

Input: BMDP-2D program for obtaining a stretched stem-and-leaf plot

```
/PROBLEM      TITLE IS 'Stem and leaf plot'.
/INPUT        VARIABLE IS 1.
              FORMAT IS '(F5.0)'.
/VARIABLE     NAME IS IQ.
/PRINT        STEM.
/END
  103
  120
  124
   .
   .
   .
  110
  100
  103
/*
//
```

Output: Stretched stem-and-leaf plot of the IQs of 70 Educational Level I boys

```
                         *
DEPTH      STEM  *  LEAVES
                         *
    2        9  *  14
   11           E  666778899
   20       10  Q  023333333
   27           *  5557799
 + 11       11  M  00000012244
   32           Q  555556677779999
   17       12  E  0001122344
    7           *  7778
    3       13  *  22
    1           *  7
                         *
DEPTH      STEM  *  LEAVES
                         *
```

In the above output the "depth" column shows the cumulative frequencies calculated from both tails of the distribution toward the stem containing the median M. These depths enable us to easily calculate the quantiles on either side of the median. The stems corresponding to the symbols E (for eighths), Q (for quartiles), M, Q and E (from top to bottom) contain respectively the 12.5th, 25th, 50th, 75th and 87.5 percentiles.

Next consider so-called box-and-whisker plots. The box-and-whisker plot of the 70 IQ scores given in Table 2.3 is shown in Figure 2.8. Such a plot is useful for depicting the locality, spread and skewness of a data set. The box indicates the median as well as the first quartile (q_1) and the third quartile (q_3). The scores above and below the box extend over the rest of the range. Although the box in Figure 2.8 is not divided into equal halves and the whiskers are not of equal length (as should be expected of a completely symmetrical distribution) the IQ values do not show any clear-cut skewness and can thus be regarded as fairly symmetrical around the median of 112. It is also possible to indicate outliers (exceptionally large or small observations) on the box-and-whisker plot. Different criteria according to which an outlier can be identified are available. A useful criterion is that of Tukey (1977) where observations larger than $q_3 + t$ or smaller than $q_1 - t$, whereby $t = 1.5(q_3 - q_1)$, are regarded as outliers. According to this criterion none of the IQ observations are very extreme.

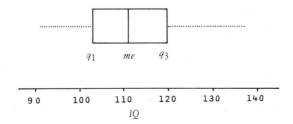

Figure 2.8. Box-and-whisker plot of 70 IQ observations.

The box-and-whisker plot furthermore offers a useful way of comparing two data sets observed in the same units with each other, with regard to locality, spread and symmetry. The next output shows the box-and-whisker plots of the scores of 28 pupils at Educational Levels II and III in the Number Series Tests (X_7 and X_{13} of Table 1.1), which were obtained through SAS PROC UNIVARIATE. The input of this program is given below.

Input: SAS PROC UNIVARIATE program for obtaining inter alia a box-and-whisker plot

```
DATA   ABILITY ;
       INPUT X7 X13 @@ ;
CARDS ;
15 19   11 21   11 15   11 12   22 19   09 17   11 21   07 13   11 13   12 15
12 09   16 19   08 10   11 16   15 15   09 12   10 19   14 21   09 13   10 12
11 15   10 17   06 12  .17 18   11 13   10 13   08 13   11 11
;
PROC   UNIVARIATE FREQ PLOT ;
TITLE  'Box and whisker plot' ;
```

Output: Box-and-whisker plot of the scores of pupils at Educational Levels II and II in the Number Series Test

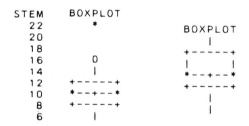

The output shows that the two data sets differ in respect of locality and spread. Both distributions indicate a degree of positive skewness. Furthermore the Educational Level II scores show two outlier values indicated by "O" (moderate) and "*" (very extreme).

4. Graphics for Bivariate Data

In the case of a bivariate data set with a small number of observations, a scatterplot provides a meaningful graphic representation (see Chapter 7). If the data set is more extensive a two-dimensional histogram can be used to represent the data. This graphic method is based on a two-dimensional frequency distribution. Table 2.7 shows a two-dimensional frequency distribution of the scores of 100 pupils in the Number Series Test at Educational

Table 2.7. Bivariate frequency distribution of the scores of 100 pupils in the Number Series Test at Educational Levels I (X_1) and II (X_7)

X_1 \ X_7	$[5; 9)$	$[9; 13)$	$[13; 17)$	$[17; 21)$	$[21; 25)$
$[22; 25)$	0	8	15	8	4
$[19; 22)$	3	19	12	4	0
$[16; 19)$	4	8	6	0	0
$[13; 16)$	3	2	4	0	0

Levels I and II (X_1 and X_7 of Table 1.1). Figure 2.9 shows the corresponding two-dimensional histogram obtained through the SAS/GRAPH PROC GCHART program.

Input: SAS/GRAPH PROC GCHART program for compilation of two dimensional histogram

```
%INCLUDE SASGRAPH;
%DEFINE (DEVICE=HP7221T,HPOS=100) ;



PROC    FORMAT ;
        VALUE I   4='(22 ; 25)'
                  3='(19 ; 22)'
                  2='(16 ; 19)'
                  1='(13 ; 16)' ;
        VALUE II 1='(5 ; 9)'
                  2='(9 ; 13)'
                  3='(13 ; 17)'
                  4='(17 ; 21)'
                  5='(21 ; 25)' ;
DATA ;
INPUT   LEVELI LEVELII SCORE @@ ;
LABEL LEVELI = 'LEVEL I' ;
LABEL LEVELII = 'LEVEL II' ;
FORMAT LEVELI   I. ;
FORMAT LEVELII II. ;
CARDS ;
4 1 0 4 2 8 4 3 15 4 4 8 4 5 4 3 1 3 3 2 19 3 3 12 3 4 4 3 5 0
2 1 4 2 2 8 2 3  6 2 4 0 2 5 0 1 1 3 1 2  2 1 3  4 1 4 0 1 5 0
;
PROC    GCHART ;
BLOCK   LEVELII/SUBGROUP=LEVELI  GROUP=LEVELI    NOLEGEND CAXIS=GREEN
                CTEXT=GREEN SUMVAR=SCORE MIDPOINTS=1 2 3 4 5 ;
PATTERN1    C=BLUE    V=L2 ;
PATTERN2    C=RED     V=L2 ;
PATTERN3    C=GREEN   V=L2 ;
PATTERN4    C=BLACK   V=L2 ;
TITLE ;
FOOTNOTE1 .F=DUPLEX .H=2 .J=L .C=BLUE   'Figure 2.9 - Bivariate histogram of sco
res of 100 pupils in Number Series Test ( X' .M=(-0.5,-0.5) .H=1 '7' .m=
(+0,+0.5) .H=2 ')' ;
```

Output:

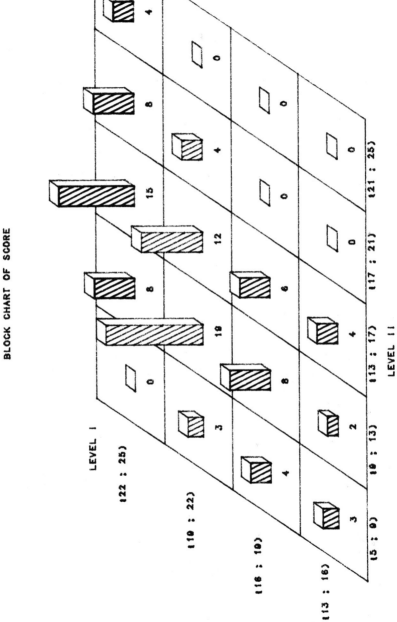

Figure 2.9. Bivariate histogram of scores of 100 pupils in Number Series Test (X_7).

Since the square bracket symbol is usually not available on the printer, the notation {5; 9) represents [5; 9), etc. . . .

5. Graphical Perception

Although many of the graphical presentations introduced in the previous sections are very striking, Cleveland and McGill (1984a) present evidence that some of these graphics do not portray the relevant information in an entirely accurate way. According to these authors there are ten elementary perceptual tasks that can be carried out to extract quantitative information from graphs. Schematically these tasks are illustrated in Figure 2.10.

Results of psychophysical experiments and empirical studies led these authors to believe that the above tasks can be ordered in the following way from the most to the least accurate.

1. Position along a common scale
2. Position along nonaligned scales
3. Length, direction, angle
4. Area
5. Volume, curvature
6. Shading, colour saturation.

In view of these results these authors propose that graphs should, as far as possible, only entail the task of judging positions along a common scale. They

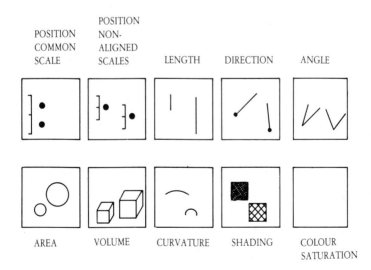

Figure 2.10. Ten elementary perceptual tasks identified by Cleveland and McGill (1984a).

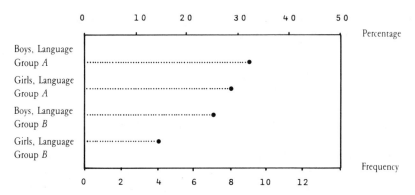

Figure 2.11. Dot chart of the composition of a sample of 28 pupils with regard to sex and language.

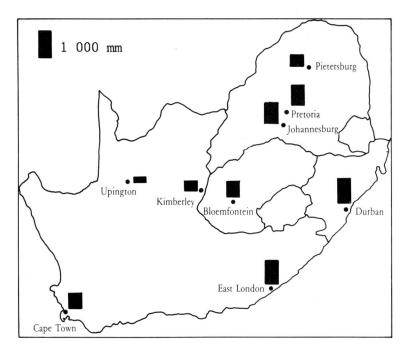

Figure 2.12. Statistical map of the annual rainfall of nine towns in the Republic of South Africa—using bars.

prefer a dot chart, for example, to a pie chart. Figure 2.11 is the dot chart corresponding to the pie chart in Figure 2.6(b).

The above ranking of the tasks also seems to suggest the use of so called framed rectangles (of equal length) other than bars of different lengths when comparing different quantities. To illustrate this consider the statistical maps

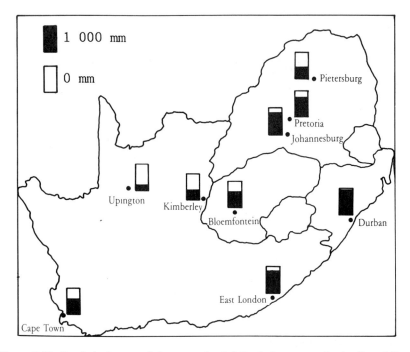

Figure 2.13. Statistical map of the annual rainfall of nine towns in the Republic of South Africa—using framed rectangles.

given in Figures 2.12 and 2.13 which give the same information by using respectively bars and framed rectangles. Obviously Figure 2.13 is easier to interpret than Figure 2.12.

As far as statistical maps are concerned, Cleveland and McGill (1984a) also maintain that the use of framed rectangles is a more unbiased method of presenting information than is the use of different shadings. This is due to the relative inaccuracy with which shading is judged and also due to the fact that large areas of dark shadings tend to catch the eye. Statistical maps can be obtained by using the SAS/GRAPH program package to which the reader is referred for further details.

Finally as a result of the fact that curvature tends to complicate accurate perception these authors also warn against the use of the well known curve-difference chart (see e.g. Playfair's (1786) chart of exports and imports from and to England from the West Indies during 1700–1780). It is much more accurate to plot only the difference curve itself.

CHAPTER 3

Graphics for Selecting a Probability Model

1. Introduction

Statistical inferential techniques can be divided into two groups, namely parametric and non-parametric (distribution-free) techniques. The difference between the two types of procedures concerns the fact that in dealing with non-parametric techniques few assumptions are made regarding the nature of the distribution of the population from which the sample is drawn. Parametric procedures, on the other hand, are based on specific knowledge of the underlying probability distribution from which the sample derives. The level of significance in parametric testing and the confidence coefficient in parametric interval estimation are therefore valid only if the assumptions made in respect of the underlying probability distribution are actually correct.

Chambers et al. (1983) present four resaons for making distributional assumptions about data. These four reasons will be summarized very briefly. Firstly, describing a set of data as a sample from a theoretical distribution achieves a valuable compactness of description of the data. Secondly, distributional assumptions can lead to some useful statistical procedures. Thirdly, such assumptions characterize the sampling distribution of statistics that may be computed during the analysis of the data and thus enable one to make inferences and probabilistic statements regarding unknown parameters of the underlying distribution. Their fourth reason is that understanding the distribution underlying the data sometimes is of value in grasping the physical mechanisms involved in generating the data.

Although the central limit theorem in the case of large samples often lends credibility to the results of inferential techniques based on the assumption of normality, the assumption that an observed sample derives from a normal population is quite frequently accepted, without justification but for the sake of convenience, in the case of small samples. In this chapter attention is paid particularly to the assessment of the normality of an observed sample. Selecting an appropriate model for a discrete variable X is also vital however, since discrete models are often used nowadays in a wide variety of situations. Discrete models are firstly considered and then continuous models.

2. Discrete Models

Assume $X_1, X_2, \ldots, X_n$ is an observed sample of a discrete variable X, and that the observations have been systematized in a frequency table as shown in Table 3.1, where n_X and $\hat{p}_X$ respectively indicate the number (frequency) and relative number of observations of size "X" in the sample.

Table 3.1. Frequency table for discrete data

X	n_X	$\hat{p}_X$
0	n_0	n_0/n
1	n_1	n_1/n
2	n_2	n_2/n
.	.	.
.	.	.
.	.	.
Total	n	1

If this data set should be analyzed by means of inferential techniques, the choice of a discrete probability model which would ensure that the data is properly described, is essential. The range of X as well as the nature of the research situation in which the data set is obtained, usually limits the number of models that can be fitted to the data to a manageable class. Four of the most widely known univariate discrete models which can be identified by means of graphical procedures, are the following:

Binomial:

$$p_X = \binom{N}{X} \pi^X (1 - \pi)^{N-X}; \qquad X = 0, 1, 2, \ldots, N; \quad 0 < \pi < 1 \cdot$$

Negative Binomial:

$$p_X = \binom{X + m - 1}{m - 1} \pi^m (1 - \pi)^X; \qquad X = 0, 1, 2, \ldots; \quad 0 < \pi < 1, \quad m > 1$$

Poisson:

$$p_X = \lambda^X e^{-\lambda}/X!; \qquad X = 0, 1, 2, \ldots; \quad \lambda > 0$$

Logseries:

$$p_X = -\theta^X/\{X \ln(1 - \theta)\}; \qquad X = 1, 2, \ldots; \quad 0 < \theta < 1.$$

Note that the logseries model can be used only for variables that can assume the nonzero values 1, 2, 3, ...

The simplest graphical procedure through which any one of the above models can be identified is that of Ord (1967, 1972). This procedure involves determining

$$\hat{U}_X = X\hat{p}_X/\hat{p}_{X-1}$$

for all observed X and plotting the points $(X, \hat{U}_X)$ for all $n_{X-1} > 5$. If these points show a linear relation $\hat{U}_X = a + bX$, one of the above models, if applicable, can be selected by using Figure 3.1.

The Ord procedure is more appropriate for large samples. A more accurate procedure for confirming Poisson or logseries models in the case of smaller samples is by means of a ln-transformation of a function of $\hat{p}_X$. According to Hoaglin (1980) a plot of the points $(X, \hat{Y}_X)$, $X = 0, 1, 2, \ldots$, where

$$\hat{Y}_X = \ln(X!\hat{p}_X)$$

showing a linear relationship, is indicative of Poisson data. Furthermore if the points $(X, \hat{V}_X)$, $X = 1, 2, 3, \ldots$, where

$$\hat{V}_X = \ln(X\hat{p}_X)$$

show a linear relationship a logseries model is applicable. Ords' procedure as well as the above two transformations may also be used in the case of the respective truncated distributions. Graphical estimators of the unknown parameters can be derived by fitting straight lines to the scatterplots.

In order to illustrate the application of the above graphical procedures, we consider the data set given in Table 3.2. These data were obtained in a bibliometrical survey conducted by the personnel of the Merensky Library of the University of Pretoria and represent a frequency table of the number of volumes of a sample of 13,076 titles from the authors' catalogue.

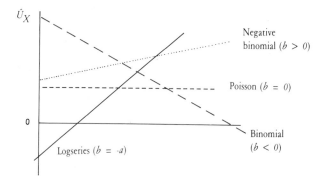

Figure 3.1. Model selection using Ord's procedure.

Table 3.2. Number of volumes of titles

Volumes (X)	Number (n_X)	$\hat{p}_X$	$\hat{U}_X$	$\hat{V}_X$
1	9860	0.7541		−0.282
2	1941	0.1484	0.394	−1.215
3	569	0.0435	0.879	−2.036
4	262	0.0201	1.842	−2.521
5	144	0.0110	2.748	−2.900
6	120	0.0092	5.000	−2.897
7	57	0.0044	3.325	−3.480
8	36	0.0027	5.053	−3.835
9	21	0.0016	5.250	−4.241
10	19	0.0015	9.048	−4.200
11	10	0.0008	5.789	−4.733
12	10	0.0008	12.000	−4.646
13	6	0.0005	7.800	−5.036
>13	21			
	13,076			

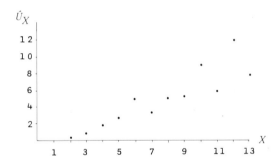

Figure 3.2. Ord representation of the number of volumes.

Figures 3.2 and 3.3 respectively show the plots of $(X, \hat{U}_X)$, $X = 2, 3, \ldots, 13$ and $(X, \hat{V}_X)$, $X = 1, 2, \ldots, 13$. From these figures it follows that the logseries model can be used to describe the observed data set although other, less well-known distributions, such as the Zeta or Yule distributions, are even more appropriate in this case.

3. Continuous Models

Statistical inferential techniques for analyzing continuous data are based mainly on the assumption that the observed data represent observations from a normal population. The effect of deviations from normality on these techniques particularly in the multivariate case, has not yet been investigated fully

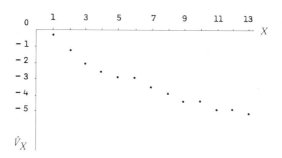

Figure 3.3. Ln (logseries) representation of the number of volumes.

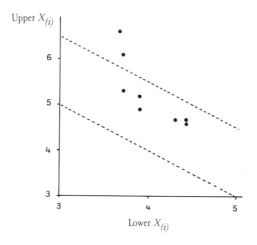

Figure 3.4. Assessing the symmetry of a distribution.

but can be serious particularly in the case of small samples. It is therefore important that this assumption be tested in respect of the data set in question, prior to applying inferential techniques which presuppose normality. Although there are analytical methods for doing this, only graphical techniques for assessing (a) univariate and (b) multivariate normality will be discussed. Before considering the problem of assessing normality we briefly discuss the simpler problem of graphically determining whether an observed data set represents a set of observations from a symmetric distribution. Indicating the ordered sample observations by $X_{(1)}, X_{(2)}, \ldots, X_{(n)}$ one now plots the points $(X_{(1)}, X_{(n)})$, $(X_{(2)}, X_{(n-1)})$, $(X_{(3)}, X_{(n-2)})$, etc. For a symmetric distribution this plot should yield a straight line with a slope of -1.

As an example consider the 17 observations of the cholesterol-level variable X_{17} for the weightlifting group in Table 1.2. The ordered values are given in Table 3.4 and the plot referred to above, in Figure 3.4.

From the above figure, despite the smallness of the sample, the two auxiliary lines with a slope of -1 serve to show the possible asymmetric nature of the distribution.

A. Normality in the Univariate Case

Distributional assumptions with regard to the data set under consideration are usually tested by making use of the so-called Q-Q (quantile-quantile) plot. A theoretical Q-Q plot is obtained by plotting the quantiles of the theoretical distribution which is being considered against the quantiles of the data. If the theoretical distribution is a close approximation to the empirical distribution one would then expect the theoretical quantilies and the quantiles of the data to nearly fall on a straight line. Vertical shifts and changes in slope do not affect the linearity of the configuration of the points in a Q-Q plot and the distributional assumptions are thus checked only according to the linearity of the plot.

The theoretical quantiles required for obtaining a Q-Q plot can usually be obtained by means of tables, approximate formulas or by using probability paper. Exact closed-form formulas can also be used but are not discussed further here. If the normal distribution is used in the theory underlying the compilation of a Q-Q plot, then a normal probability plot is obtained.

If the sample of observations X_1, X_2, ..., X_n is not too large (say $n < 50$) a normal probability plot is constructed as follows:

Indicate the ordered sample observations by $X_{(1)}$, $X_{(2)}$, ..., $X_{(n)}$. Subsequently assume that $U_{(1)}$, $U_{(2)}$, ..., $U_{(n)}$ are the means of n ordered observations obtained from a $n(0; 1)$ population through repeated sampling. These $U_{(i)}$'s, $i = 1, 2, ..., n$ are called "rankits" and are tabled in *Biometrika Tables for Statisticians*, Vol. 1 of E.S. Pearson and H.O. Hartley (1966), Table 28 which is given below. The graphical representation of the $U_{(i)}$'s against the $X_{(i)}$'s is done on ordinary graph paper and is called a rankit plot. This plot should show a linear relation if X_1, X_2, ..., X_n do in fact represent observations from a normal popluation.

The normal probability plot is now illustrated by checking whether the 17 observations of variable X_{17} (cholesterol level) in Table 1.2 for the weightlifting group actually derive from a normal population. The ordered observations $X_{(i)}$, together with the corresponding rankits $U_{(i)}$, are given in Table 3.4.

The representation of $U_{(i)}$ against $X_{(i)}$ is given in Figure 3.5 and shows some deviations from linearity indicating that the assumption of normality may perhaps be unrealistic. There are various techniques according to which it can be assessed whether the points in such a probability plot are indicative of a linear relationship. Mage (1982) inter alia discusses a graphical technique for assessing the linearity of such a relationship.

If normal probability paper is available Z_i-values can be plotted against the $X_{(i)}$, where Z_i is usually chosen from the following possibilities:

Table 3.3. Mean positions of ranked normal deviates (rankits)

i \ n	2	3	4	5	6	7	8	9	10	11	12
1	0·564	0·846	1·029	1·163	1·267	1·352	1·424	1·485	1·539	1·586	1·629
2		·000	0·297	0·495	0·642	0·757	0·852	0·932	1·001	1·062	1·116
3				·000	·202	·353	·473	·572	0·656	0·729	0·793
4						·000	·153	·275	·376	·462	·537
5								0·000	0·123	0·225	0·312
6										·000	·103

i \ n	13	14	15	16	17	18	19	20	21	22	23	24	25
1	1·668	1·703	1·736	1·766	1·794	1·820	1·844	1·867	1·89	1·91	1·93	1·95	1·97
2	1·164	1·208	1·248	1·285	1·319	1·350	1·380	1·408	1·43	1·46	1·48	1·50	1·52
3	0·850	0·901	0·948	0·990	1·029	1·066	1·099	1·131	1·16	1·19	1·21	1·24	1·26
4	·603	·662	·715	·763	0·807	0·848	0·886	0·921	0·95	0·98	1·01	1·04	1·07
5	0·388	0·456	0·516	0·570	0·619	0·665	0·707	0·745	0·78	0·82	0·85	0·88	0·91
6	·190	·267	·335	·396	·451	·502	·548	·590	·63	·67	·70	·73	·76
7	·000	·088	·165	·234	·295	·351	·402	·448	·49	·53	·57	·60	·64
8			·000	·077	·146	·208	·264	·315	·36	·41	·45	·48	·52
9					·000	·069	·131	·187	·24	·29	·33	·37	·41
10							0·000	0·062	0·12	0·17	0·22	0·26	0·30
11									·00	·06	·11	·16	·20
12											·00	·05	·10
13													·00

i \ n	26	28	30	32	34	36	38	40	42	44	46	48	50
1	1·98	2·01	2·04	2·07	2·09	2·12	2·14	2·16	2·18	2·20	2·22	2·23	2·25
2	1·54	1·58	1·62	1·65	1·68	1·70	1·73	1·75	1·78	1·80	1·82	1·84	1·85
3	1·29	1·33	1·36	1·40	1·43	1·46	1·49	1·52	1·54	1·57	1·59	1·61	1·63
4	1·09	1·14	1·18	1·22	1·25	1·28	1·32	1·34	1·37	1·40	1·42	1·44	1·46
5	0·93	0·98	1·03	1·07	1·11	1·14	1·17	1·20	1·23	1·26	1·28	1·31	1·33
6	·79	·85	0·89	0·94	0·98	1·02	1·05	1·08	1·11	1·14	1·17	1·19	1·22
7	·67	·73	·78	·82	·87	0·91	0·94	0·98	1·01	1·04	1·07	1·09	1·12
8	·55	·61	·67	·72	·76	·81	·85	·88	0·91	0·95	0·98	1·00	1·03
9	·44	·51	·57	·62	·67	·71	·75	·79	·83	·86	·89	0·92	0·95
10	0·34	0·41	0·47	0·53	0·58	0·63	0·67	0·71	0·75	0·78	0·81	0·84	0·87
11	·24	·32	·38	·44	·50	·54	·59	·63	·67	·71	·74	·77	·80
12	·14	·22	·29	·36	·41	·47	·51	·56	·60	·64	·67	·70	·74
13	·05	·13	·21	·28	·34	·39	·44	·49	·53	·57	·60	·64	·67
14		·04	·12	·20	·26	·32	·37	·42	·46	·50	·54	·58	·61
15			0·04	0·12	0·18	0·24	0·30	0·35	0·40	0·44	0·48	0·52	0·55
16				·04	·11	·17	·23	·28	·33	·38	·42	·46	·49
17					·04	·10	·16	·22	·27	·32	·36	·40	·44
18						·03	·10	·16	·21	·26	·30	·34	·38
19							·03	·09	·15	·20	·25	·29	·33
20								0·03	0·09	0·14	0·19	0·24	0·28
21									·03	·09	·14	·18	·23
22										·03	·08	·13	·18
23											·03	·08	·13
24												·03	·07
25													0·03

The table gives the expectation $\xi(i\,|\,n)$ of the ith largest observation in a sample of n normal deviates, the unit being the population standard deviation.

The columns for n from 21 to 50 are taken from *Statistical Tables for Biological, Agricultural and Medical Research* (Fisher & Yates, 1963, Table XX) by permission of the authors and the publishers, Messrs Oliver and Boyd.

Table 3.4. Ordered observations and rankits for X_{17} (cholesterol level) for the weightlifting group in Table 1.2

Observation:	3.66	3.72	3.76	3.9	3.91	4.33	4.43	4.44	4.49
Rankit :	−1.794	−1.319	−1.029	−0.807	−0.619	−0.451	−0.295	−0.146	0.000
Observation:	4.57	4.66	4.70	4.88	5.23	5.33	6.10	6.57	
Rankit :	0.146	0.295	0.451	0.619	0.807	1.029	1.319	1.794	

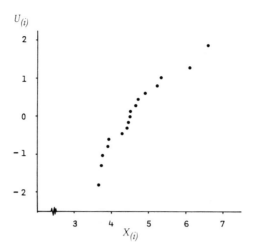

Figure 3.5. Plot of rankits against the ordered observations of variable X_{17} in Table 1.2.

$$i/(n + 1)$$
$$(i - 0.5)/n$$
$$Z_i = (i - 0.3)/(n + 0.4) \qquad i = 1, 2, \ldots, n.$$
$$(i - 0.375)/(n + 0.25)$$
$$(i - 0.33)/(n + 0.33)$$

Normal probability paper is specially prepared graph paper in which one of the scales (the vertical scale in Figure 3.6) represents a linearisation of the cumulative normal distribution.

The first two formulas for Z_i are usually used although the BMDP programme series uses the last formula in the 5D program. The motivation for choosing the above formulas is as follows: If the unit area under the normal curve is divided into n equal sections it can be expected that if normality holds, one observation from the total of n observations will fall in each section. For $Z_i = (i - 0.5)/n$ this consequently means that $X_{(i)}$ is plotted against the midpoint $(i - 0.5)/n$ of the cumulative area of the ith section. As before, a linear probability plot confirms that the sample represents observations from a normal population. The ordered observations for the variable X_{17} (see Table 1.2) and the corresponding values of $Z_i = (i - 0.5)/n$ are given in Table 3.5.

In Figure 3.6 the values of Z_i are represented against the $X_{(i)}$ on normal probability paper. The points in this probability plot again suggest that the X_i evidently do not represent observations drawn from a normal population. In order to gain experience in evaluating such plots, actual normal values for different sample sizes can be read from normal tables and plotted against Z_i,

Table 3.5. Ordered observations, $X_{(i)}$ and corresponding values of $Z_i = (i - 0.5)/n$ for variable X_{17} in Table 1.2

$X_{(i)}$	3.66	3.72	3.76	3.90	3.91	4.33	4.43	4.44	4.49	4.57	4.66	4.70
Z_i	0.03	0.09	0.15	0.21	0.26	0.32	0.38	0.44	0.50	0.56	0.62	0.68

$X_{(i)}$	4.88	5.23	5.33	6.10	6.57
Z_i	0.74	0.79	0.85	0.91	0.97

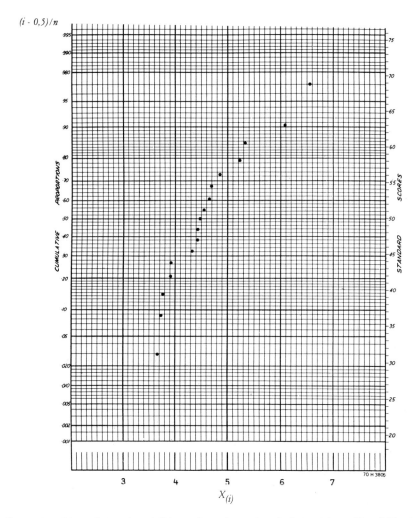

Figure 3.6. Plot of $Z_i = (i - 0.5)/n$ against the ordered observations $X_{(i)}$ of X_{17} in Table 1.2.

Table 3.6. Cumulative frequency table of the IQ-values of 70 boys at Educational Level II

Class limit	F	F/(n + 1)
95	2	0.03
100	11	0.15
105	20	0.28
110	27	0.38
115	38	0.54
120	53	0.75
125	63	0.89
130	67	0.94
135	69	0.97
140	70	0.99

whereby one then becomes attuned to the deviations of linearity which may occur in a sample that does in fact derive from a normal population.

For samples larger than say, 50 the data can be grouped in a frequency table (see Table 2.1). The cumulative frequency, which is expressed as a proportion of $(n + 1)$ where n represents the sample size, is then plotted against X on normal probability paper where X usually assumes the values of the upper class limits. If this plot takes the form of a straight line, the data are representative of observations deriving from a normal population.

Table 3.6 represents the cumulative frequency table obtained by using the BMDP-2D stem-and-leaf representation of the data of Table 2.3. F indicates the cumulative frequency and in Figure 3.7 $F/(n + 1)$ is plotted against the IQ-values.

From Figure 3.7 it appears from the linear nature of the plot that these 70 observations represent a sample from a normal population. If it is suspected that the sample derives from a normal population with a zero mean, a so-called half normal plot can be drawn. In this case the absolute values of the observations X_i can be arranged in sequence and $0.5 + (i - 0.5)/n$ is plotted against the absolute values of $X_{(i)}$ on normal probability paper. This representation is used particularly in residual analysis (see Chapter 7) where the absolute values of the residuals are used. Such a half normal plot can also be obtained by specifying the HALFNORM option in the BMDP-5D program which is illustrated shortly.

A normal probability plot as shown in Figure 3.6 or 3.7 may also be used to estimate the unknown mean μ and standard deviation σ of the normal population from which the sample derives. The value on the horizontal scale which corresponds to 0.50 on the vertical scale, is the estimated value for the median or the mean μ. The difference between the values on the horizontal

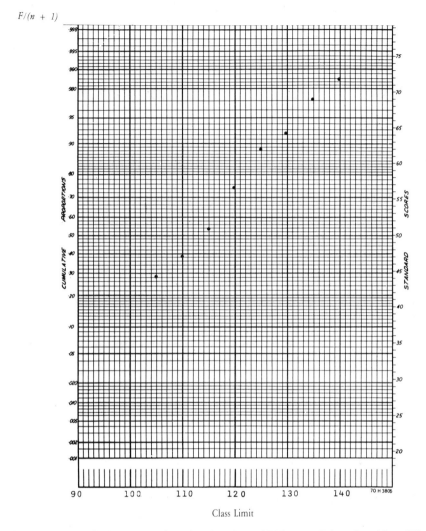

Figure 3.7. Plot of $F/(n + 1)$ against the IQ-values of 70 boys at Educational Level II.

scale which correspond to respectively 0.84 and 0.50 on the vertical scale, is an estimate of σ. Application of this procedure in Figure 3.7 produces the following estimates for μ and σ

$$\hat{\mu} = 112, \qquad \hat{\sigma} = 9.$$

As an example of a computer representation of a probability plot the BMDP-5D program is subsequently used to check for normality in the case of the 17 observations for X_{17} for the group of weightlifters. The plot given by BMDP-5D must be compared to the probability plot in Figure 3.5.

Input: BMDP-5D program for the compilation of a normal probability plot

```
/PROBLEM      TITLE IS 'TEST FOR NORMALITY ON 17 OBSERVATIONS FOR X17'.
/INPUT        VARIABLE IS 1.
              FORMAT = '(F5.2)'.
              CASES ARE 17.
/VARIABLE     NAME IS X17.
/PLOT         TYPE=NORM.
              SIZE=30,20.
/END
 4.44
 4.88
 4.33
 3.66
 4.57
 3.90
 3.91
 4.43
 5.33
 3.76
 6.57
 4.66
 4.70
 6.10
 5.23
 4.49
 3.72
/*
//
```

The size of the figure is determined by means of the SIZE statement in the PLOT paragraph.

Output:

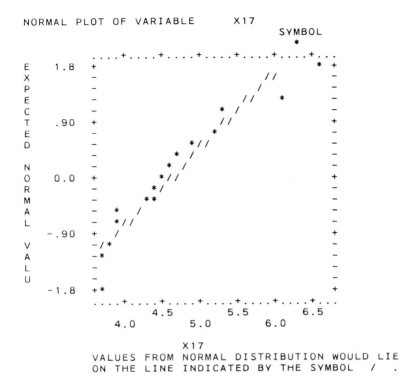

```
NORMAL  PLOT  OF  VARIABLE        X17
                                         SYMBOL
                                            *

          . . . . + . . . . + . . . . + . . . . + . . . . + . . . . + . . .
   E    1.8   +                                          *  +
   X          −                              / /            −
   P          −                            /                −
   E          −                        / /      *           −
   C          −                  *  /                        −
   T    .90   +                / /                          +
   E          −            *                                −
   D          −         * / /                               −
              −       *  /                                  −
   N          −     *   /                                   −
   O    0.0   +     * / /                                   +
   R          −    * /                                      −
   M          −    * *                                      −
   A          −   *   /                                     −
   L          −   * / /                                     −
       −.90   +   /                                        +
   V          − / *                                         −
   A          − *                                           −
   L          −                                             −
   U          −                                             −
       −1.8   + *                                          +
          . . . . + . . . . + . . . . + . . . . + . . . . + . . . . + . . .
                4.5           5.5           6.5
            4.0           5.0           6.0
```

```
                    X17
        VALUES  FROM  NORMAL  DISTRIBUTION  WOULD  LIE
        ON  THE  LINE  INDICATED  BY  THE  SYMBOL    /    .
```

The identification of deviatory patterns which are indicative of non-normality of the data, is very important when the assumption of normality with respect to the data set under consideration, is being assessed. From the following figures the most important deviatory patterns can be identified. The vertical axes represent the normal quantiles and the ordered observations are given on the horizontal axes.

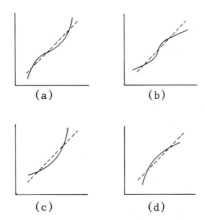

(a)   (b)

(c)   (d)

These plots are indicative of: (a) The distribution has shorter (lighter) tails than the normal distribution, (b) the distribution has longer (heavier) tails than the normal distribution, (c) the distribution is negatively skew and (d) the distribution is positively skew.

B. Normality in the Multivariate Case

A simple graphical procedure for assessing multivariate normality is described next.

Assume X_i is the ith p-variate vector of observations (for the Ability Test Data set in Table 1.1, $i = 1, 2, ..., 28$ and $p = 21$), $\bar{X}$ is the vector of sample means and S the sample covariance matrix. Next define the generalized Mahalanobis distance d_i of each point, X_i from the sample mean vector $\bar{X}$ as:

$$d_i = \sqrt{(X_i - \bar{X})'S^{-1}(X_i - \bar{X})}, \qquad i = 1, 2, ..., n.$$

If the sample derives from a multivariate normal population the d_i^2-values are approximately distributed as $\chi^2(p)$ for large samples. A χ^2-probability plot of the $\chi^2(p)$-percentiles against the ordered d_i^2-values should then yield a straight line through the origin.

The χ^2-percentiles (t_i) which are plotted against the $d_{(i)}^2$ are obtained as follows:

Solve the following equation for r_i:

$$[1/\Gamma(p/2)] \int_0^{r_i} (X/2)^{p/2-1} \exp(-X/2)\, dX = Z_i$$

where $Z_i = (i - 0.5)/n; i = 1, 2, ..., n$.

The χ^2-percentiles are then given by $t_i = 2r_i; i = 1, 2, ..., n$. These percentiles can be obtained by means of the SAS function GAMINV as illustrated in the SAS program given next or can be read from tables of gamma quantiles given by Wilk, Gnanadesikan and Huyett (1962).

Next consider the 28 observations of the variables X_1 to X_6 in the data set of Table 1.1. The assessment of multivariate normality is illustrated on the basis of these data. The distances d_i^2, the χ^2-percentiles and the resulting probability plot are all obtained by means of the following SAS PROC MATRIX and SAS PROC PLOT programs.

Input: SAS PROC MATRIX and SAS PROC PLOT program for obtaining a χ^2-probability plot

```
PROC    MATRIX ;
A =
  19 21 21 18 20 21/
  21 20 15 24 22 18/
  18 19 16 18 18 23/
  18 23 10 18 16 16/
```

```
           24 24 19 20 23 24/
           19 19 23 21 23 23/
           21 20 19 21 21 23/
           21 20 21 20 16 22/
           19 20 19 22 18 21/
           19 23 22 22 16 25/
           17 13 08 18 13 18/
           21 22 22 15 23 23/
           18 18 17 16 15 22/
           13 18 21 16 17 15/
           17 13 17 20 22 19/
           18 12 09 09 15 17/
           22 15 24 17 15 20/
           18 17 18 18 13 18/
           17 15 14 14 12 13/
           16 20 17 13 15 16/
           24 21 22 21 21 25/
           23 23 21 22 16 21/
           22 22 21 24 18 24/
           22 17 19 19 21 20/
           20 23 23 22 22 24/
           22 17 21 17 17 22/
           21 18 20 23 21 22/
           21 22 19 20 18 17 ;
* Calculation of covariance matrix and sample mean ;
           N = NROW(A) ;
           NVAR = NCOL(A) ;
* Set elements of SCOV and XBAR equal to zero ;
           SCOV = J(NVAR,NVAR,0.) ;
           XBAR = VECDIAG(SCOV) ;
* Calculate mean vector ;
DO I = 1 TO N ;
           XBAR = XBAR + A(I,)' ;
END ;
           XBAR = XBAR #/ N ;
* Calculate sample covariance matrix and inverse ;
DO I = 1 TO N ;
           Z = A(I,)' - XBAR ;
           SCOV = SCOV + Z*Z' ;
END ;
           SCOV = SCOV #/ (N-1) ;
           INV_S = INV(SCOV) ;
* Calculate the Mahanalobis distances D(I) ;
           D = J(N,1,0) ;
DO I = 1 TO N ;
           ZI = A(I,)' - XBAR ;
           ZIS = ZI' * INV_S ;
           D(I,) = ZIS * ZI ;
END ;
* Order distances in ascending order ;
           R = RANK(D) ;
           TEMP = D ;
           D(R,) = TEMP ;
* Calculate percentiles by means of SAS 'GAMINV' function ;
           P = J(N,1,0) ;
DO I = 1 TO N ;
           Z = (I-0.5) #/ N ;
           GAMM = NVAR #/ 2 ;
           P(I,) = 2 * GAMINV(Z,GAMM) ;
END ;
           T = D || P ;
OUTPUT T OUT=NEW ( RENAME = (COL1=MAHANAL COL2=PERCENT) ) ;
PROC   PRINT DATA=NEW ;
       VAR MAHANAL PERCENT ;
TITLE Ordered distances and corresponding percentiles ;
* Plot Chi-squared percentiles against ordered Mahanalobis distances ;
PROC   PLOT DATA=NEW ;
       PLOT PERCENT * MAHANAL = '*' / VPOS=25 HPOS=40 ;
TITLE  Plot of percentiles against ordered Mahanalobis distances ;
```

Output:

(i) *Ordered distances d_i^2 and corresponding χ^2-percentiles*

```
OBS  MAHANAL  PERCENT
  1   1.2205   1.0860
  2   1.5951   1.6830
  3   1.6594   2.0954
  4   1.8796   2.4411
  5   2.9143   2.7518
  6   3.0826   3.0419
  7   3.1456   3.3191
  8   3.2777   3.5886
  9   3.4918   3.8540
 10   3.5842   4.1180
 11   3.6363   4.3830
 12   3.9917   4.6510
 13   4.3076   4.9242
 14   4.3844   5.2046
 15   5.3403   5.4943
 16   5.3730   5.7960
 17   6.0795   6.1124
 18   6.3360   6.4472
 19   6.5052   6.8047
 20   7.7495   7.1907
 21   8.4399   7.6129
 22   8.7908   8.0822
 23   9.7150   8.6148
 24   9.9767   9.2364
 25  10.4436   9.9917
 26  10.5829  10.9707
 27  12.0570  12.4024
 28  12.4398  15.3275
```

(ii) χ^2-Probability plot

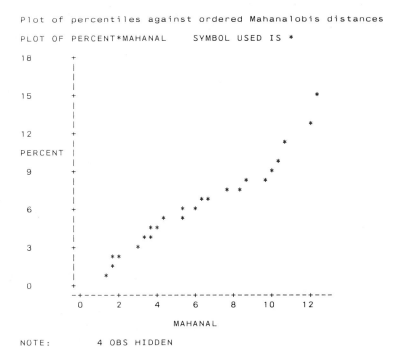

```
Plot of percentiles against ordered Mahanalobis distances

PLOT OF PERCENT*MAHANAL       SYMBOL USED IS *

  18        +
            |
            |
            |
  15        +                                                    *
            |
            |                                                *
            |
  12        +
            |                                            *
PERCENT     |
            |                                        *
   9        +                                    *
            |                                *   *
            |                            *   *
            |                        * *
   6        +                      *   *
            |                    *   *
            |                *   *
            |              * *
   3        +              * *
            |          * *
            |          *
            |      *
   0        +
            -+-----+-----+-----+-----+-----+-----+---
             0     2     4     6     8    10    12

                          MAHANAL

NOTE:        4 OBS HIDDEN
```

Although the χ^2-probability plot in the output above indicates a slightly curvilinear trend the deviation from linearity is not drastic and the assumption that the sample of X_1, X_2, ..., X_6-scores derives from a six-variate normal distribution would therefore be acceptable.

4. General

In Section 2 the Zeta and Yule distributions were mentioned in connection with the fitting of a model to the data obtained from a bibliometrical survey. These two distributions are discussed more fully in Johnson and Kotz (1969).

The Q-Q plot introduced in Section 3 is discussed in great detail by Chambers et al. (1983) who consider various other aspects such as Q-Q plots in the case of distributions with an unknown shape parameter. Wilk and Gnanadesikan (1968) also present a comprehensive description of the Q-Q plot and its applications.

In the section on assessing multivariate normality only a simple graphical procedure is presented. Gnanadesikan (1977) discusses the assessment of multivariate normality in more detail both from the standpoint of evaluating the similarity of marginal distributions and of preserving the multivariate nature of the data. He discusses both analytical and graphical methods.

CHAPTER 4

Visual Representation of Multivariate Data

1. Introduction

During the past two or three decades a considerable number of highly expressive visual techniques have been developed for representing multivariate data in two dimensions. Some of these techniques, such as Andrews' curves and Chernoff faces, have captured the imagination of the research community (particularly the non-statisticians) and are currently enjoying widespread interest. Some ten representations will be discussed in this chapter and illustrated on the basis of simple examples.

Certain of these techniques have serious shortcomings and are difficult to apply to large data sets unless the relevant computer programs are available, but even so, are noteworthy on account of the insight that they offer into multivariate data sets that would otherwise be difficult to interpret.

2. "Scatterplots" in More Than Two Dimensions

A three-dimensional point or vector (X_1, X_2, X_3) can be represented in two dimensions at the two-dimensional point (X_1, X_2) by choosing the size of the point proportional to X_3. To illustrate this we consider the following example. The first three scores of the five pupils given in Table 4.1 are respectively $(16, 16, 19)$; $(14, 17, 15)$; $(24, 23, 21)$; $(18, 17, 16)$ and $(18, 11, 9)$.

In Figure 4.1 these scores are represented in "three dimensions" by means of a triple scatterplot. The score with the smallest X_3 value is represented by the smallest circle whereas the score with the largest X_3 value is represented by the largest circle.

Alternatively an extension of the scatter plot to four, five, or six dimensions

Table 4.1. Scores of five pupils in six ability
tests (a random smaple from the complete
Ability Test Data set of 2800 pupils—see
Section 4, Chapter 1)

Pupil	X_1	X_2	X_3	X_4	X_5	X_6
1	16	16	19	21	20	23
2	14	17	15	22	18	22
3	24	23	21	24	20	23
4	18	17	16	15	20	19
5	18	11	9	18	7	14

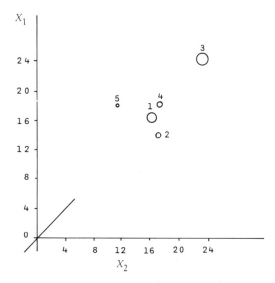

Figure 4.1. Triple scatterplot of scores of five pupils in three ability tests.

can be obtained by drawing rectangular lines from each point (X_1, X_2) respectively proportional to the value of X_3, X_4, X_5 and X_6. Figure 4.2 represents
the scores of the five pupils as given in Table 4.1 in the six ability tests.

From Figures 4.1 and 4.2 it follows that the scores of pupils 1 and 2 and to
a lesser degree pupil 4 are fairly homogeneous, whereas the scores of pupils 3
and 5 differ totally from these three.

3. Profiles

The most natural and the simplest way of representing a p-dimensional
observation is by using p vertical columns or bars. The bars are placed next
to each other with the heights respectively proportional to the p observations.

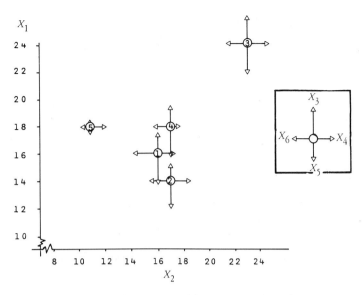

Figure 4.2. Plot of the scores of five pupils in six ability tests.

Such a bar diagram is constructed for every observation vector. Alternatively the midpoints of the bars can be joined to give a more continuous appearance to the profiles. The total appearance of the profiles is naturally very dependent on the ordering of the variables.

Figure 4.3 shows the profiles of the five pupils in respect of the six scores

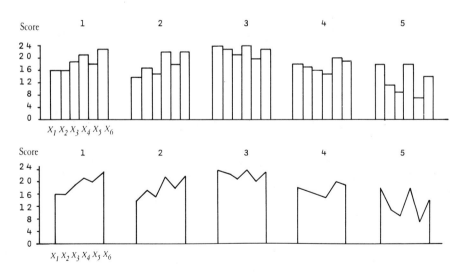

Figure 4.3. Two alternative profile representations of five pupils' scores in six ability tests.

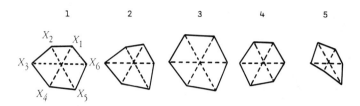

Figure 4.4. Star representations of five pupils' scores in six ability tests.

$X_1, X_2, \ldots, X_6$ (see Table 4.1). Except for a degree of correspondence between pupils 1 and 2, the other profiles show very little similarity. In general, negative values are represented by bars or points below the horizontal line.

4. Star Representations

Stars or polygons represent the p measurements of a case on equally spaced radii extending from the centre of a circle. The measurements are then linked, to form a star. Such a star can now be formed for every observation vector. In stars the values of the p variables are arranged in a circle as opposed to the parallel arrangement of profiles. Stars provide a more finished representation than profiles, and the fact that the starting points correspond to the end points facilitates comparison between the cases. However if p is large, star representations are often not very elucidating. This display, also known as circle diagrams or snowflakes, will be illustrated on the basis of the same five data vectors used in the previous sections. Figure 4.4 shows these star representations which can also be obtained by utilising the SAS/GRAPH program package referred to in Chapter 2. In this example the size of the star can be regarded as a measurement of the total achievement in the Ability Tests. Consequently pupils 3 and 5, generally speaking, respectively fared the best and the worst. In general, in the case of negative values, the values of the particular variables can be made positive through a zeropoint transformation (adding a constant to each value of the variable).

5. Glyphs

Anderson (1960) proposed that the measurements of each of the p variables in a specific case be represented as rays from a circle. These representations are known as glyphs and differ in two respects from the star representations discussed in the previous section. Firstly, all the rays are drawn from the top of the circle to the outside and secondly, the end points of the rays are not linked as was the case with the stars. Such a glyph is then drawn for every observation vector. Anderson provides the following guidelines to help in observing differences between glyphs:

○ Seven variables can be represented at the most.
○ Divide the range of every variable into three categories, namely low, average and high, and make the lengths of the corresponding rays zero, one unit and two units respectively. Alternatively quartiles (or even deciles) may be used to divide the range into more intervals.
○ Variables that are highly correlated should be associated as far as possible with adjacent rays.
○ If two or more divergent types of cases (for instance men and women) are observed, the information can be represented in a glyph by distinguishing between the circles (for instance colored for men and blank for women).
○ $(p + 2)$-variate observation vectors can be represented by representing the p-variate glyphs in a scatter plot at position (X_{p+1}, X_{p+2}) (see Figure 4.2).

Figure 4.5(a) shows the glyphs of the scores of the five pupils in the six ability tests.

According to Anderson all the information in a glyph can be pooled by means of an index which is calculated as the sum of the lengths of all the rays. If a ray can have a length of only 0, 1 or 2, the index will vary between 0 and $2p$. If there are a large number of important variables (more than seven), the four or five showing the highest linear relationship can be transformed into a single index as described above. The indices together with the values of the other variables can then be used in glyph representations. The index value (I) of each of the five glyphs in Figure 4.5(a) appears below each glyph. Since X_2,

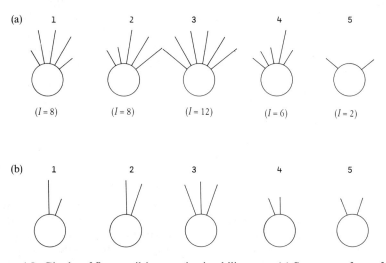

Figure 4.5. Glyphs of five pupils' scores in six ability tests. (a) Sequence of rays from left to right $(X_1, X_2, X_3, X_5, X_6, X_4)$. (b) Sequence of rays from left to right (X_1, I, X_4).

X_3, X_5, and X_6 are fairly highly correlated, these four variables can be reduced to a single variable by calculating an appropriate index value. These index values are calculated as follows:

Pupil	X_2	Recoding X_3	X_5	X_6	Index I	Classification
1	1	2	2	2	7	High
2	1	1	2	2	6	High
3	2	2	2	2	8	High
4	1	1	2	1	5	Average
5	0	0	0	0	0	Low

In Figure 4.5(b) the glyphs for the five pupils for (X_1, I, X_4) are given in sequence from left to right.

6. Boxes

Hartigan (1975) recommended that boxes be used for representing multivariate data. If $p = 3$, each observation vector is represented by one box with the length, width and height corresponding to the X_1, X_2 and X_3 observations. (All negative values are made positive by the addition of an appropriate constant.) For $p > 3$ a single side of the box can be used to represent more than one variable by dividing the side into segments. In such cases it is important to represent related (highly correlated) variables by means of segments of the same side, since the human eye is more adept at identifying differences between the length of sides than between segments of a specific side. Grouping together highly correlated variables on one side of a box reduces the variation between segment lengths of a side for individual boxes.

In Figure 4.6(a) the first three ability test scores of the five pupils are represented as boxes whereas Figure 4.6(b) represents the six ability test scores. The vertical sides of the boxes are divided into four segments which correspond to the highly correlated tests X_2, X_3, X_5 and X_6.

The disadvantage of this method is that it becomes unsuitable if p is large (larger, say, than 8) and furthermore, it is strongly influenced by the ordering of the variables in respect of the sides and the segments of a side.

7. Andrews' Curves

Andrews (1972) proposed the use of harmonic functions for presenting multivariate data points. He introduced the function

$$f_{\mathbf{X}}(t) = X_1/\sqrt{2} + X_2 \sin t + X_3 \cos t + X_4 \sin 2t + X_5 \cos 2t + \cdots$$

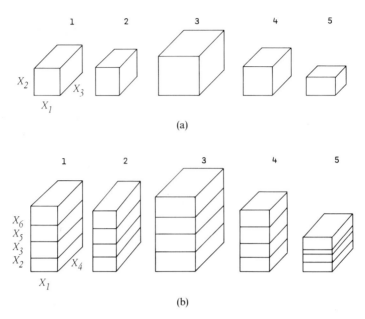

Figure 4.6(a). Box representation of five pupils' scores in three ability tests.   (b). Box representation of five pupils' scores in six ability tests.

for $-\pi < t < \pi$, as a two-dimensional representation of the observation vector $\mathbf{X} = (X_1, X_2, \ldots, X_p)'$.

Calculating function values for different values of t can easily be done with simple computer facilities. Accurate graphical representation is a laborious process however, unless computer graph plotters are available.

If the functions of the n, p-variate observation vectors are plotted on the same axial system, the cases can be compared with one another. If n is large however (larger, say, than 10), the eye finds it difficult to follow the course of such a large number of individual functions and to draw comparisons. In this case n separate Andrews' curves can be plotted, each one on its own axial system and then compared with one another.

Two important properties of the Andrews' function $f_{\mathbf{X}}(t)$ are:

○ The function representation preserves means. This implies that the function of the mean vector of the observed vectors $\mathbf{X}_1, \mathbf{X}_2, \ldots, \mathbf{X}_n$ is the pointwise mean of the functions $f_{\mathbf{X}_1}(t), f_{\mathbf{X}_2}(t), \ldots, f_{\mathbf{X}_n}(t)$.
○ The function representation preserves distances. It can be proved that the distance between two functions $f_{\mathbf{X}_1}(t)$ and $f_{\mathbf{X}_2}(t)$ measured by

$$\int_{-\pi}^{\pi} [f_{\mathbf{X}_1}(t) - f_{\mathbf{X}_2}(t)]^2 \, dt$$

is proportional to the Euclidean distance (see Chapter 5) between the observation vectors $\mathbf{X}_1$ and $\mathbf{X}_2$.

These two properties give rise to the fact that Andrews' curves are useful in clustering observation points in homogeneous groups or to compare individual functions with the mean function.

Although other harmonic functions could also be used to represent multivariate data and although many of these functions also have some of the qualities of the Andrews' function $f_X(t)$, the last mentioned function is preferred on account of its excellent mathematical and statistical characteristics. Under the assumption that the p variables are independent and each is distributed according to the normal distribution, Andrews derived significance tests for testing whether an individual observation vector X_0 differs significantly from the hypothetical population mean μ. A test for differences between $f_{X_0}(t)$ and $f_\mu(t)$ for a particular value of t is also described by Andrews (1972).

In the graphical representation of Andrews' curves low frequencies are observed more easily than high frequencies, for which reason Andrews recommends that the most important variables should be associated with the low frequencies. More satisfactory results are obtained when Andrews' curves are drawn from the principal components (see Chapter 6) rather than from the raw data. In such cases X_1, X_2, ... should be chosen respectively as the first, second, ... principal components of the data set.

Chernoff (1973) recommended that the Andrews' curves should be represented by a circle rather than a horizontal scale for t. This gives rise to representations of which the starting points and end points correspond, and different observation vectors should therefore not be drawn on one axial system.

In order to illustrate the use of Andrews' curves the six Ability Test scores of the five pupils in the six ability tests considered earlier, are drawn as five curves by means of the following SAS/GRAPH program. The five functions are respectively.

$$f_1(t) = 16/\sqrt{2} + 16\sin t + 19\cos t + 21\sin 2t + 20\cos 2t, + 23\sin 3t$$

$$f_2(t) = 14/\sqrt{2} + 17\sin t + 15\cos t + 22\sin 2t + 18\cos 2t + 22\sin 3t$$

$$f_3(t) = 24/\sqrt{2} + 23\sin t + 21\cos t + 24\sin 2t + 20\cos 2t + 23\sin 3t$$

$$f_4(t) = 18/\sqrt{2} + 17\sin t + 16\cos t + 15\sin 2t + 20\cos 2t + 19\sin 3t$$

$$f_5(t) = 18/\sqrt{2} + 11\sin t + 9\cos t + 18\sin 2t + 7\cos 2t + 14\sin 3t.$$

Input: SAS/GRAPH PROC GPLOT program for compilation of Andrews' curves

```
DATA    ABILITY ;
        SQRT2=SQRT(2) ;
        KEEP F1-F5 T ;
        PIE=3.1415926535898 ;
        INCR=2*PIE/100 ;
DO T=-PIE TO PIE BY INCR ;
    F1=16/SQRT2+16*SIN(T)+19*COS(T)+21*SIN(2*T)+20*COS(2*T)+23*SIN(3*T) ;
    F2=14/SQRT2+17*SIN(T)+15*COS(T)+22*SIN(2*T)+18*COS(2*T)+22*SIN(3*T) ;
    F3=24/SQRT2+23*SIN(T)+21*COS(T)+24*SIN(2*T)+20*COS(2*T)+23*SIN(3*T) ;
    F4=18/SQRT2+17*SIN(T)+16*COS(T)+15*SIN(2*T)+20*COS(2*T)+19*SIN(3*T) ;
    F5=18/SQRT2+11*SIN(T)+ 9*COS(T)+18*SIN(2*T)+ 7*COS(2*T)+14*SIN(3*T) ;
LABEL  F1='00'X
       T ='00'X ;
OUTPUT ;
END ;  * T-LOOP ;
PROC    GPLOT ;
PLOT    (F1-F5)*T / OVERLAY   HAXIS= -3 -2 -1 0 1 2 3
                    NOLEGEND  VAXIS= -40 TO 110 BY 10 ;
SYMBOL1 L=1 I=JOIN C=RED ;
SYMBOL2 L=2 I=JOIN C=GREEN ;
SYMBOL3 L=3 I=JOIN C=BLUE ;
SYMBOL4 L=4 I=JOIN C=ORANGE ;
SYMBOL5 L=5 I=JOIN C=BLACK ;
TITLE1 .F=DUPLEX .H=2 'Andrews Curves of 5 Pupils scores in 6 Ability Te
sts' ;
```

Output:

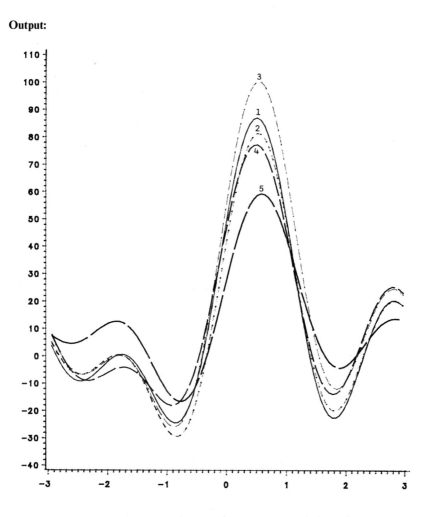

Andrews curves of 5 pupils scores in 6 Ability Tests

All the graphical methods discussed thus far in this chapter lead to the same conclusions regarding the achievements of the five pupils in tests X_1 to X_6: pupils 1 and 2 (and perhaps 4) obtained reasonably similar scores whereas the scores obtained by pupils 3 and 5 differ totally from one another, and from those of pupils 1, 2, and 4.

8. Chernoff Faces

Considerable interest is currently being shown in the representation of multi-variate data points by means of faces (or rather charicatures of the human face). This technique was originally introduced by Chernoff (1973) with a view to representing data of 18 dimensions at the most. Using a face makes sense because:

○ widely divergent facial features are shown, each of which can be associated with the different variables, and
○ all of us are accustomed to distinguishing accurately between faces with different features.

The procedure is very simple and can be adjusted or adapted at will to suit the needs of a particular data set. As an example:

X_1 can be associated with the size of the mouth,
X_2 with the size of the nose,
X_3 with the shape of the chin, etc.

In the case of measurements on human beings it is possible to relate the total facial expression to the meaning of the data point, or in fact to convert measurements of a person's personality traits into a face that matches the latter. Measurements signifying that a person is friendly by nature can be represented by a laughing face for instance, and depression by a surly face.

In a case where fewer variables are observed than the features which can be varied, the least important features can be left constant. If certain variables are related (for instance measurements of the same type of characteristic in personality testing), these variables can be associated with a specific facial area (say the eyes and eyebrows). Cases can therefore also be compared with one another in respect of such a subcollection of variables, by specifically noting this facial area.

A major drawback of Chernoff faces is that the subjective assigning of features to variables has a marked effect on the eventual shape of the face. If the aim of the data analysis is merely the classification of faces in groups, the permutations of the assigning of features was shown by Chernoff and Rizvi (1975) to cause an error percentage of as high as 25%. This means that classifying two faces as "fairly similar" is greatly influenced by the assigning of variables to specific features. The most important variables should therefore be assigned to the most prominent features. A problem that arises however, is that the choice of prominent features is also a subjective matter. If a computer program is used for obtaining the representations, considerable experimenting with the program and permutations of variable feature combinations is required before the Chernoff procedure can be used with confidence. This is particularly the case if p is large (larger, say, than 8).

Whereas Chernoff's (1973) method is able to deal with 18 variables at the most, Flury and Riedwyl (1981) have expanded the procedure to provide for a maximum of 36 variables. This they have achieved by changing symmetrical features (e.g. nose, mouth, hair) to assymmetrical, which means that the left side of the mouth may look different from the right side, etc.

It is virtually impossible to represent Chernoff faces without using a computer program. A graph plotter is also useful although not indispensable. The program of Tidmore and Turner (1977) which will be briefly discussed on the basis of the example below, can deal with a maximum of 12 variables and "draws" the faces by means of an ordinary line printer.

If $p > 12$ a principal component analysis (see Chapter 6) can be carried out on the data and the first (major) 12 principal components can be used for presenting the faces. In such cases faces can be compared globally only, since individual features no longer represent original variables.

This program uses the following feature–variable combinations:

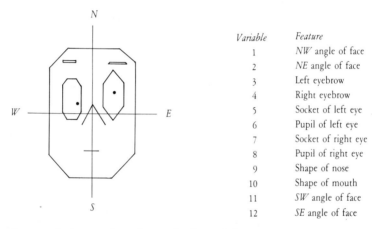

Variable	Feature
1	NW angle of face
2	NE angle of face
3	Left eyebrow
4	Right eyebrow
5	Socket of left eye
6	Pupil of left eye
7	Socket of right eye
8	Pupil of right eye
9	Shape of nose
10	Shape of mouth
11	SW angle of face
12	SE angle of face

The above rule governing the assigning of features to variables may be amended to accommodate other variable-feature combinations.

In order to illustrate this program we use the mean monthly rainfall of a representative sample of 15 randomly chosen rainfall districts in South Africa. These data are given in Table 4.2.

The twelve months, January to December were assigned to features 1 to 12 in accordance with the feature–variable combinations mentioned above. The Chernoff faces for these 15 districts are shown in Figure 4.7. These faces show that widely different rainfall patterns exist. For example the near identical faces 2 and 3 represent winter rainfall areas whereas, in contrast, faces 23 and 29 represent predominantly summer rainfall areas.

Suppose a Chernoff face representation of the 18 ability tests, X_1 to X_{18}, is required for each of the 28 pupils (see Table 1.1). In a situation like this where more than 12 variables are available and all are important in describing the profile of a data vector, one can use the first 12 principal component scores of an individual to represent the original variables. Since the first principal

Table 4.2. Mean monthly rainfall (in mm) of 15 districts in South Africa (Source: SA Statistics 1982. Central Statistical Service)

District	J	F	M	A	M	J	J	A	S	O	N	D
2	3.9	8.6	8.9	23.3	41.0	60.4	44.9	45.7	23.7	17.5	13.4	5.9
3	7.9	10.6	11.8	29.7	56.4	75.0	62.6	59.5	35.2	25.3	15.3	8.9
11	78.5	72.6	86.7	67.0	73.9	57.0	62.8	78.8	90.6	88.5	88.9	78.2
17	14.1	21.9	28.1	17.5	15.4	9.2	11.1	10.5	12.3	14.0	19.8	18.0
23	96.7	91.2	121.9	66.4	43.2	25.7	30.5	33.8	66.8	85.4	105.4	91.5
29	118.3	111.8	110.9	47.7	32.4	21.4	20.6	24.1	41.5	73.7	105.2	112.6
33	153.4	149.5	119.2	60.0	20.9	16.3	15.7	12.1	35.2	65.0	119.1	138.8
39	22.8	32.4	45.5	27.2	18.6	8.1	10.0	9.3	11.5	15.5	25.5	23.4
42	74.5	73.4	81.2	40.1	25.6	11.2	12.7	16.3	25.7	37.8	63.6	67.7
50	87.3	67.7	47.0	26.8	7.8	6.7	5.4	2.2	7.2	28.8	55.4	77.5
54	28.9	41.5	49.9	33.0	17.3	7.3	7.9	7.5	10.1	18.2	27.8	22.5
60	116.3	91.9	84.9	48.6	27.1	11.5	12.1	11.1	30.7	71.1	95.6	10.5
75	118.0	92.3	80.1	45.0	21.9	7.8	7.4	6.4	19.7	63.6	109.4	112.9
79	55.5	62.9	61.0	36.0	16.6	7.0	3.4	4.9	5.2	17.7	30.2	38.8
83	88.6	73.3	76.1	45.8	22.2	8.4	9.0	7.0	12.5	46.4	65.6	71.8

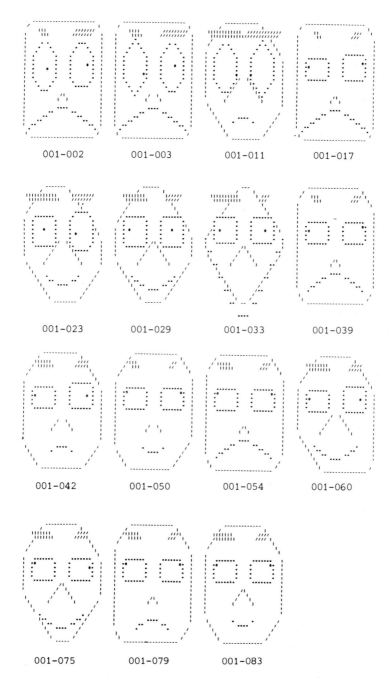

Figure 4.7. Chernoff faces of mean monthly rainfall of 15 districts in South Africa.

component Y_1 contains the most information, the second principal component Y_2 the second most, etc. it makes sense to assign Y_1 to the most important feature, Y_2 to the second most important feature, etc. Although such an assigning of features is somewhat subjective, it was decided to do it as follows:

Y_1 —Shape of mouth
Y_2 —Shape of nose
Y_3 —SW angle of face ⎱
Y_4 —SE angle of face ⎰   Shape of chin
Y_5 —NW angle of face ⎱
Y_6 —NE angle of face ⎰   Forehead
Y_7 —Socket of left eye
Y_8 —Socket of right eye
Y_9 —Left eyebrow
Y_{10}—Right eyebrow
Y_{11}—Pupil of left eye
Y_{12}—Pupil of right eye

In Figure 4.8 the symbol 001-001 for example designates the first pupil from group 1 (Language group A, boys). In the data set given in Table 1.1 this therefore refers to pupil number 3.

In studying the faces of the 28 pupils it is possible to detect relatively similar faces (e.g. 003-001 and 004-001) and widely different faces (e.g. 001-003 and 003-003). The two last mentioned faces correspond respectively to pupils number 5 and 16 (see Table 1.1). In Chapter 6 different scaling methods show these two pupils' ability test profiles to be widely different.

Differences between the four groups can be enhanced further by obtaining a prototype face for each group through using the mean scores of $Y_1, Y_2, \ldots,$

Language Group A, boys

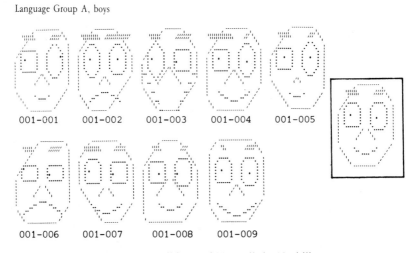

Figure 4.8. Chernoff faces of 28 pupils in 18 ability tests.

Language Group A, girls

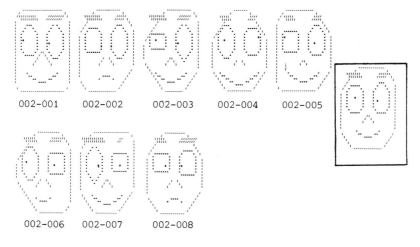

002-001 002-002 002-003 002-004 002-005

002-006 002-007 002-008

Language Group B, boys

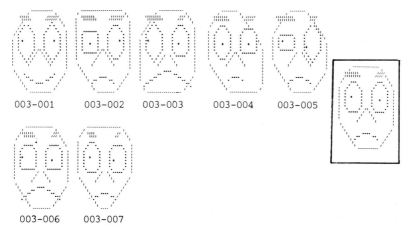

003-001 003-002 003-003 003-004 003-005

003-006 003-007

Language Group B, girls

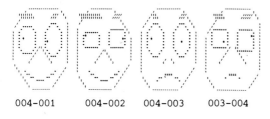

004-001 004-002 004-003 003-004

Figure 4.8 (continued)

Y_{12} for each group. These faces are indicated on the right-hand side of each group in Figure 4.8. Using the well-known fact that the first, most important, component Y_1 is an overall summarizing measure of the original variables (X_1 to X_{18} in this case), differences in the appearance of the mouths of the prototype faces are of special interest. The achievement of the boys of Language Group B in the data set seems therefore different (worse) than that of the other 3 groups.

The input for the 32 faces in Figure 4.8 is as follows.

Input: FACES program for representing Chernoff faces

```
32 12                                                                                Section 1
10 9 11 12 1 2 5 7 3 4 6 8                                                           Section 2
 1  1  -40  -14    28    28    94  -64  -53    61  -13    -6  -18 -106               Section 3
 1  2 -370 -104   -18   209    20   72    58    13    79   183   -2  -29                .
 1  3  530  177    54   144   142   26   140  -74  -41    48  -31   56                  .
 1  4  245   49   -98    27    39   13   -44    56   137  -67  -15   -9                  .
 1  5  -27 -204   114   -60    91  -32   -53    80    -8  -57    10  -26                  .
 1  6 -424   17   -84   -33   147 -140   -96 -149  -56    25  -89   43                  .
 1  7  256  -67   195   -13   -48   50   -31  -89    42    18  -19   51                  .
 1  8   81  230  -220    -3    60  105   -53  -21  -72  -56    68   52                  .
 1  9  391  -37   -71    24  -127  -47   -31  -18 -149     5   -3  -15                  .
 2  1  196   63  -182  -227  -152 -108    37    76    53    93  -57   52                  .
 2  2 -133  -95  -132    44    29    2   -13    52    41     3   157   75                  .
 2  3  211 -127   137     1   135 -112   -97    95    23    23    -0   47                  .
 2  4  306 -149    20   -24    93   93    56    -5  -45  -21     9 -100                  .
 2  5  273 -242  -220    -5  -132   69   -54    12  -42  -14  -70   15                  .
 2  6    8  -92   169  -213    90  -25    61  -71  -18    29  -27   19                  .
 2  7  186 -132   -12    24   -69 -153   125 -117   135  -81    98   -1                  .
 2  8  -78  -87   -18    26     3   69   -61  -35  -56    93    89   51                  .
 3  1  230  164    30    12    41    8   111   150  -46    25  -19  -46                  .
 3  2 -207  161   135     8  -225  -92   -50    34     4    68    39  -63                  .
 3  3 -649   39  -178   -29    66  -18   160    25    51  -52  -61    3                  .
 3  4 -150   52   137  -262   -68  189    71  -43     2   -7    47  -15                  .
 3  5 -215  130   125   -13    32  125  -115    95    79  -38  -23   63                  .
 3  6 -482   35    87    38   -91   39   -32  -53  -19  -43  -56    5                  .
 3  7 -240   -4   165   202  -145  -34    42    -4  -75  -92   -1   30                  .
 4  1  225  173    88    55   -29  -83    16    41    -3  -39  -25   50                  .
 4  2  287   51   -86    80   -42   94   -96 -113   129  -10  -74  -86                  .
 4  3 -272 -173   -91    27   -25   37    64    60  -93  -37  -28   -9                  .
 4  4 -138  187   -75   -66    70  -83   -63  -61  -40     8    99 -109                  .
 5  1   71    5   -11    36    47   -2   -18  -16    -9    10  -11    2                  .
 5  2  121 -108   -30   -47    -0  -21     7     1    11    16    25   20                  .
 5  3 -245   82    72    -6   -55   31    27    29    -1  -20  -11   -3                  .
 5  4   26   59   -41    24    -7   -9   -20  -18    -2  -20    -7  -39               Section 3
```

The input consists of the following three subsections.

Section 1: n, p (Free Format)

n = total number of cases
p = total number of variables

Section 2: $i_1, i_2, \ldots, i_p$ (Free Format)

$i_j, j = 1, 2, \ldots, p$ = sequence of characteristic number

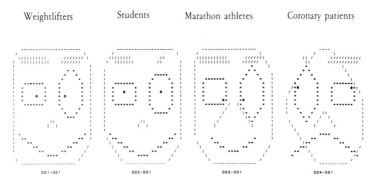

Figure 4.9. Chernoff faces of means of the four groups in the Cholesterol/Fitness data set.

Section 3: $a, b, X_1, X_2, \ldots, X_p$ (Free format)

   a = group number
   b = number of case in specific group
   $X_1, X_2, \ldots, X_p$ = values of variables (free format)

   Chernoff faces can be used to relay a message in a simple and informative way. Figure 4.9, for example, shows the "average faces" of the four groups mentioned in the Fitness/Cholesterol data (see Table 1.2) in respect of the eleven cholesterol variables. Variable 12 (SE angle of face was held constant). It is obvious that the coronary patients differ from the other three groups.

9. General

For further applications of star representations see e.g., Siegel et al. (1972) and Herman and Montroll (1972). Kulkarni and Paranjape (1984) suggest some alternative harmonic functions which can also be used in representing multivariate data.

   A considerable number of studies have already been conducted with a view to determining which of the graphical representations discussed in Sections 2 to 8 are best able to bring to the fore differences between observations (see, for example, Wang (1978) pp. 123–141). It appears that Andrews' curves normally fare the best, Chernoff faces second best and profiles the worst, with not much to choose from between stars, glyphs and boxes. It is interesting to note that research has shown that a graphic representation of the first two principal components or other multidimensional scaling procedures in two dimensions (see Chapter 6) are much more adept at highlighting differences between observations than any of the visual methods discussed in this chapter. It cannot be argued away however, that the use of visual representations can liven up an otherwise dull data representation while at the same time offering a considerable amount of information not readily available in a data matrix.

Although all the representations discussed in this chapter are intended for quantitative observations (usually continuous), categorical data of an ordinal nature can also be represented by means of these representations, provided the categories are coded ordinally (for example 0, 1, 2, etc.).

CHAPTER 5

Cluster Analysis

1. Introduction

As in the previous chapter this chapter mainly deals with multivariate data. A multivariate data set can conveniently be expressed in the form of a matrix as follows:

$$
\mathbf{X} = \begin{bmatrix} X_{11} & X_{12} & \cdots & X_{1p} \\ X_{21} & X_{22} & \cdots & X_{2p} \\ \vdots & \vdots & & \vdots \\ X_{n1} & X_{n2} & \cdots & X_{np} \end{bmatrix} = \begin{bmatrix} \mathbf{X}'_1 \\ \mathbf{X}'_2 \\ \vdots \\ \mathbf{X}'_n \end{bmatrix}.
$$

The rows of $\mathbf{X}$ therefore, represent the n different cases while the columns represent the p variables.

Graphical techniques such as those described in the previous chapter give indications of differences between observed cases. If the cases seem very heterogeneous in respect of the characteristics that are measured it is often necessary and advantageous to group them in a systematic way into fairly homogeneous groups or clusters. The procedures by means of which such clusters are formed is known as cluster analysis. Although two-dimensional representations are used to aid the researcher in this clustering of cases, these graphs are essentially an ordering of the n cases in one dimension according to their mutual proximity or distance.

Two approaches can be followed in cluster analysis, namely the probability and deterministic approaches. The former will be discussed briefly in Section 2. Since the degree of distance and similarity between observation vectors plays a central role in deterministic approaches some important measures of distance and similarity are introduced in Section 3. The deterministic approach to cluster analysis usually gives rise to hierarchical procedures and will be discussed in Sections 4 and 5. Although the aim of cluster analysis is usually to group observation vectors into clusters, it can also be very useful in identifying related variables. Cluster analysis of variables is looked into in Section 8. A few lesser known graphical representations which could be fruitfully used in the field of cluster analysis, are discussed in Section 6, 7, and 10.

2. The Probability Approach

If the observation vectors $\mathbf{X}_1, \mathbf{X}_2, \ldots, \mathbf{X}_n$ are independent, the probability approach entails choosing clusters $(T_1, T_2, \ldots, T_g)$ which will maximize the likelihood function for every possible grouping of observations into g clusters. The grouping which produces the overall maximum of the likelihood function then constitutes the optimal grouping.

If every cluster or subsample of observation vectors can be regarded as coming from a multivariate normal population, the probability approach reduces to that choice of clusters $(T_1, T_2, \ldots, T_g)$ which minimizes

$$|\mathbf{W}| = \left| \sum_{k=1}^{g} \sum_{T_k} (\mathbf{X}_i - \bar{\mathbf{X}}_k)(\mathbf{X}_i - \bar{\mathbf{X}}_k)' \right|$$

($|\mathbf{W}|$ represents the determinant of $\mathbf{W}$). This method is not often used in practice since it requires a considerable amount of computer time and is based on somewhat restricting assumptions in respect of normality and independence of the observations. Consequently, deterministic (non-probability) procedures are used much more frequently in cluster analysis.

3. Measures of Distance and Similarity

The clustering methods which will be discussed in this chapter are either based on a symmetric distance matrix $\mathbf{D}$ or on a symmetric similarity matrix $\mathbf{C}$. The matrices $\mathbf{D}$ and $\mathbf{C}$ can be observed directly but are more usually calculated from the data matrix $\mathbf{X}$.

Distances

A numerical distance function $d(\mathbf{X}_i, \mathbf{X}_j)$ of pairs of p-dimensional points of the data set $\mathbf{X}$ is said to be a metric if it satisfies the following conditions:

(i) $d(\mathbf{X}_i, \mathbf{X}_j) \geq 0$; $d(\mathbf{X}_i, \mathbf{X}_j) = 0$ if $\mathbf{X}_i = \mathbf{X}_j$;
(ii) $d(\mathbf{X}_i, \mathbf{X}_j) = d(\mathbf{X}_j, \mathbf{X}_i)$; (symmetry)
(iii) $d(\mathbf{X}_i, \mathbf{X}_k) + d(\mathbf{X}_j, \mathbf{X}_k) \geq d(\mathbf{X}_i, \mathbf{X}_j)$ (triangular inequality).

For convenience we will indicate $d(\mathbf{X}_i, \mathbf{X}_j)$ by d_{ij} and arrange all the distances in a symmetric matrix $\mathbf{D}$.

$$\mathbf{D} = \begin{bmatrix} d_{11} & d_{12} & \cdots & d_{1n} \\ d_{21} & d_{22} & \cdots & d_{2n} \\ \vdots & \vdots & & \vdots \\ d_{n1} & d_{n2} & \cdots & d_{nn} \end{bmatrix}.$$

The most general distance metric is the Minkowski metric

$$d_{ij} = \left\{ \sum_{l=1}^{p} |X_{il} - X_{jl}|^k \right\}^{1/k}.$$

The following two special cases are very important:

(i) The Euclidean metric ($k = 2$)

$$d_{ij} = \sqrt{\sum_{l=1}^{p} (X_{il} - X_{jl})^2}.$$

(ii) The "City block" or L_1-metric ($k = 1$)

$$d_{ij} = \sum_{l=1}^{p} |X_{il} - X_{jl}|.$$

A generalization of the Euclidean metric is the weighted Euclidean metric given by

$$d_{ij} = \sqrt{\sum_{l=1}^{p} w_l (X_{il} - X_{jl})^2}.$$

Two standardized distance metrics are used very frequently. The first, known as the Pearson or χ^2 distance metric is a special case of the weighted Euclidean metric with $w_l = 1/s_l^2$. Therefore,

$$d_{ij} = \sqrt{\sum_{l=1}^{p} (X_{il} - X_{jl})^2 / s_l^2}$$

where s_l is the calculated standard deviation of variable X_l. The other standardized metric, which is known as the Mahalanobis distance measure (see Chapter 3), is defined by

$$d_{ij} = \sqrt{(\mathbf{X}_i - \mathbf{X}_j)'\mathbf{S}^{-1}(\mathbf{X}_i - \mathbf{X}_j)}$$

where $\mathbf{S}$ is the estimated covariance matrix of the population from which the observations $\mathbf{X}_1, \mathbf{X}_2, \ldots, \mathbf{X}_n$ were sampled.

If the information on n cases is contained in an ($I \times J$) contingency table, a matrix of distances between the row or column categories can be obtained by applying any of the above measures to the frequencies. The Euclidean distance between row categories i and i' is, for example, defined by

$$d_{ij} = \sqrt{\sum_{j=1}^{J} (f_{ij} - f_{i'j})^2}.$$

Some researchers have derived non-metric measures of distance which may be applied usefully in a variety of problems. Two of these are the Calhoun distance and the Lance and Williams non-metric distance measure.

Finally consider n groups (cases) which are each classified into p categories. The rows of the data matrix $\mathbf{X}$ are then respectively the observed proportions from populations of sizes $N_1, N_2, \ldots, N_n$. Each row can therefore be regarded as a realization of a $(p - 1)$-variate multinomial variable (an example of the above is the observed ratio of persons in n countries belonging to blood groups A_1, A_2, B, and O). Alternative distance measures for such qualitative type of data can be used and are discussed in Mardia, Kent, and Bibby (1979).

Similarities

The usual assumptions concerning the elements of the similarity matrix $\mathbf{C}$ are

(i) $c_{ij} \geq 0$, $c_{ij} = 1$ for $i = j$;
(ii) $c_{ij} = c_{ji}$ (symmetry).

In the case of quantitative data, distances are used more frequently than similarities to represent the degree of association between cases. Although the well-known Pearson correlation coefficient can be used to measure the similarity between cases for quantitative data, its use has been criticized by many authors on a number of different grounds.

For qualitative data, similarity or association coefficients are more frequently used to measure the relationship between two cases or individuals. Assume a binary data matrix $\mathbf{X} = \{X_{ij}\}$; $i = 1, 2, \ldots, n$ and $j = 1, 2, \ldots, p$ where:

$$X_{ij} = \begin{cases} 1 \text{ if characteristic } j \text{ is present in case } i \\ 0 \text{ otherwise.} \end{cases}$$

In order to define similarity coefficients consider some data for two cases:

Variable (Characteristic)

	1	2	3	4	5	6	7	8	9	10
Case 1:	0	1	1	0	0	0	0	1	1	0
Case 2:	0	1	1	1	0	1	0	1	0	0

The above can be summarized in a two-way association table.

Case 1

		1	0	
Case 2	1	3	2	5
	0	1	4	5
		4	6	10

In general for cases i and j we have the following table:

Case i

		1	0	
Case j	1	a	b	$a + b$
	0	c	d	$c + d$
		$a + c$	$b + d$	p

Similarity coefficients which assume values in the interval 0 to 1 can be defined in a variety of ways in terms of the constants a, b, c and d. Some of the more widely used are:

○ Russell and Rao

$$c_{ij} = a/p.$$

○ Sokal and Michener (simple matching)

$$c_{ij} = (a + d)/p.$$

○ Jaccard

$$c_{ij} = a/(a + b + c).$$

The above three measures possess reasonably useful probabilistic interpretations.

Gower (1971) has defined a general similarity coefficient which can be used for qualitative and quantitative data. This similarity coefficient can also be used with data containing a mixture of qualitative and quantitative variables.

Transforming Distances to Similarities

Cases that are a great distance from each other indicate little similarity whereas small distances indicate large similarity. A further difference between these two types of measures is that whilst similarities assume values between 0 and 1, distance measures can assume any positive value. A distance d_{ij} can be transformed into a similarity by the transformation

$$c_{ij} = (1 + d_{ij})^{-1}.$$

The reverse process is not so obvious because of the triangular inequality which must be satisfied by a distance metric. Assuming that the similarity matrix $\mathbf{C} = \{c_{ij}\}$ is positive semi-definite

$$d_{ij} = \sqrt{c_{ii} - 2c_{ij} + c_{jj}}$$

is the standard transformation from $\mathbf{C}$ to $\mathbf{D}$ and results in an Euclidean distance matrix. In the special case where $c_{ii} = c_{jj} = 1$ this transformation simplifies to:

$$d_{ij} = \sqrt{2(1 - c_{ij})}.$$

4. Hierarchical Cluster Analysis

In a hierarchical cluster analysis the observation vectors (cases) are generally grouped together on the basis of their mutual distances. Taking n clusters as the point of departure (with every case forming a cluster) the cases are grouped together in a stepwise manner until all the observation vectors have been grouped together in one cluster.

This procedure possesses a hierarchical characteristic in that a case allocated to a certain grouping cannot be reallocated during a later grouping. Eventually the composition of each cluster is determined by a so-called threshold value distance. The stages of the analysis are usually represented graphically by means of a tree diagram.

In order to illustrate a hierarchical cluster analysis, we consider the data in Table 4.1 and carry out a so-called single linkage analysis. Table 5.1 represents Pearson's distance matrix for the five pupils.

Table 5.2 indicates the eleven stages in this analysis. The composition of the clusters after each step appears in the final column. The procedure is as follows:

○ Each pupil is regarded as a separate cluster.
○ Group the two pupils together having the smallest mutual distance; in this case, pupils 1 and 2. Only four clusters remain.
○ The next shortest distance is d_{14}. Pupil 4 is therefore grouped with cluster $(1, 2)$ in Step 3.
○ This procedure is continued by checking all the distances in the order of shortest to longest and grouping the clusters together until all the pupils are grouped in one cluster.
○ Represent the steps of the grouping graphically in a tree diagram.

Figure 5.1 gives the so-called dendrogram for these five observations which can be constructed by means of the information contained in Table 5.2. If a threshold value of $d_0 = 6$, for instance, is chosen as the distance between clusters (see Figure 5.1), the five observations consist of three clusters, namely $(1, 2, 4); (3); (5)$. Figure 5.2 illustrates another representation of a tree diagram, namely the cluster map in which clusters consisting of a single observation

Table 5.1. Pearson's distance matrix for the data of Table 4.1

Pupil no.	1	2	3	4	5
1	0				
2	1.38	0			
3	8.17	11.35	0		
4	4.73	9.27	13.29	0	
5	17.94	14.25	31.01	12.01	0

Table 5.2. Stages in the single linkage
cluster analysis of five pupils in six
ability tests

Step no.	Distances	Cluster
1	0	(1); (2); (3); (4); (5)
2	$d_{12} = 1.38$	(1, 2); (3); (4); (5)
3	$d_{14} = 4.73$	(1, 2, 4); (3); (5)
4	$d_{13} = 8.17$	(1, 2, 3, 4); (5)
5	$d_{24} = 9.27$	(1, 2, 3, 4); (5)
6	$d_{23} = 11.35$	(1, 2, 3, 4); (5)
7	$d_{45} = 12.01$	(1, 2, 3, 4, 5)
8	$d_{34} = 13.29$	(1, 2, 3, 4, 5)
9	$d_{25} = 14.25$	(1, 2, 3, 4, 5)
10	$d_{15} = 17.94$	(1, 2, 3, 4, 5)
11	$d_{35} = 31.01$	(1, 2, 3, 4, 5)

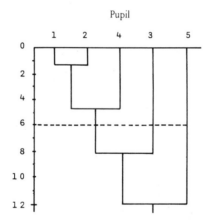

Figure 5.1. Dendrogram of five pupils in six ability tests.

are indicated by "∗" and larger clusters by an unbroken line. The different
clusters can easily be read from this figure for the different values of g, the
number of clusters. Thus it is obvious that for $g = 3$ the composition of the
clusters are (1, 2, 4); (3); (5).

The above example illustrates the hierarchical method of single linkage
cluster analysis. According to this method the "distance" between two clusters
is defined as the minimum distance between any two p-dimensional points
(cases) of the two clusters. If there are $(g + 1)$ clusters, the two clusters which
according to this criterion are the "closest" to each other are grouped together,
whereby g clusters are formed. A large variety of hierarchical methods of

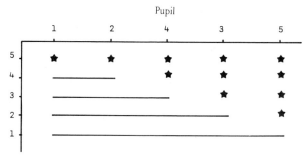

Figure 5.2. Cluster map of five pupils in six ability tests.

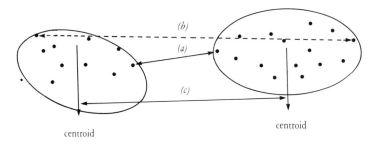

Figure 5.3. Criteria for distance between clusters. (a) Single linkage (nearest neighbor), (b) complete linkage (furthest neighbor), and (c) centroid linkage.

cluster analysis are available. These methods differ in respect of the criteria used for identifying the two clusters situated "closest" to each other. Figure 5.3 shows a two-dimensional schematic representation of three possible cluster distance criteria which can be used in hierarchical cluster analysis.

From this figure it appears that the complete linkage criterion, as opposed to the single linkage criterion, defines "cluster distance" as the maximum distance between two points of the two clusters. The centroid linkage criterion, on the other hand, defines cluster distance as the distance between the two centroids (for example, means or medians) of the two clusters. A fourth cluster distance criterion which is also used frequently, is the so-called average linkage criterion where cluster distance is defined as the arithmetic mean of all the possible distances between two points of the two clusters.

Ward (1963) proposed a fifth cluster distance criterion. His measure entails that at any stage of a cluster analysis the loss of information which results from the grouping of cases into clusters should be measured by the total sum of squared deviations of every observed vector from the mean of the cluster to which it belongs. For every pair of clusters (at each stage) the increase in the "error sum of squares" due to this fusion is calculated and the pair associated with a minimum increase is then combined.

The five hierarchical methods discussed above are called agglomerative methods. The opposite way, that is, starting with only one cluster which is divided into two and proceeding until n clusters are formed, leads to the so-called divisive methods. These methods are also hierarchical in nature but will not be discussed here.

The order of clustering (and consequently also the dendrograms and cluster maps) for the five pupils which were obtained by means of all five the above methods, are identical. However, the cluster distances which are indicated on the vertical axes in the dendrograms (see Figure 5.1) differ on account of the differing definitions of cluster distances.

Advantages and Disadvantages of Hierarchical Cluster Analysis

Apart from the fact that hierarchical methods of cluster analysis are acceptable from a mathematical point of view, these various hierarchical procedures also have the following advantages:

○ No assumptions are made in respect of the probability distribution of the p-dimensional population from which the n observations are obtained.
○ The techniques are relatively simple and for small values of n and p the calculations can even be carried out on a pocket calculator.
○ The cluster groupings remain unchanged irrespective of any monotone transformation of the distance matrix **D**.

Hierarchical procedures have certain disadvantages however, the most important of which undoubtedly is the so-called chain effect. These chains are a result of the tendency of relatively isolated observation points, which occur between natural clusters, to link these clusters together. This effect is in fact caused by the hierarchical nature of the analysis.

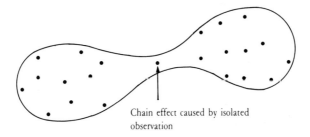

Chain effect caused by isolated observation

The figure above indicates a two-dimensional representation of the chain effect. Wishart (1969) proposed a modified single linkage technique known as mode analysis in order to counteract this negative effect in hierarchical cluster analysis. This procedure involves the initial removal of isolated points prior to carrying out a single linkage analysis.

The somewhat subjective choice of a threshold value through which the clusters are eventually compiled can be regarded as a disadvantage in hierarchical cluster analysis. We should bear in mind however, that hierarchical cluster analysis is an exploratory deterministic analysis and is used only for identifying relatively homogeneous groups of observations in a single data set and rarely as a classical inferential technique. Meaningful clusterings for a given value of g, the number of clusters, can often be obtained by rather using the cluster map (see Figure 5.2).

Cluster Analysis Based on a Similarity Matrix

If a similarity matrix is available for a data set, hierarchical cluster analysis can be carried out by

(a) grouping clusters with maximal intercluster similarity together at each step, or
(b) first transforming the similarity matrix C into a distance matrix D (see Section 3) and then carrying out a cluster analysis in the usual way.

Cluster Analysis in the Case of Contingency Tables

In an $(I \times J)$ contingency table hierarchical cluster analysis can be carried out on the I row categories or J column categories separately by starting out from the matrix of distances which, according to Section 3, is calculated for the rows or columns, respectively. The clusters of row categories thus formed can then be regarded as the new row categories. Similarly the number of column categories can also be reduced to fewer categories. Pooling the categories of a contingency table is frequently necessary in the case of relatively small values of n in order to make the χ^2-approximations of test statistics more valid. For further information regarding cluster analysis in the case of contingency tables see Cohen (1980).

5. Computer Programs for Hierarchical Cluster Analysis

The five best known hierarchical cluster analysis techniques were briefly presented in the preceding section. If it is taken into account that these analyses can be based on any one of the distance (or similarity) matrices considered in Section 3, it is obvious that a wide variety of analyses are possible. Two programs, namely, the SAS PROC CLUSTER/TREE program and the BMDP-2M program will subsequently be used to carry out a cluster analysis on the data of Table 1.1, variables X_1 to X_{18}.

Input: SAS PROC CLUSTER/TREE program for cluster analysis

```
DATA    ABILITY ;
INPUT   X1-X18 ;
NUMBER = _N_ ;
CARDS ;
19 21 21 18 20 21 15 14 15 13 15 16 19 19 19 20 17 17
21 20 15 24 22 18 11 18 16 19 14 17 21 15 17 18 18 19
18 19 16 18 18 23 11 13 13 15 11 11 15 18 13 15 18 13
                .
                .
                .
22 17 21 17 17 22 10 14 16 16 13 08 13 18 21 12 13 15
21 18 20 23 21 22 08 15 09 17 11 13 13 20 20 21 15 20
21 22 19 20 18 17 11 15 12 14 11 10 11 13 14 14 15 14
;
PROC    CLUSTER STD METHOD=A OUTTREE=TREE ;
ID      NUMBER ;
VAR     X1-X18 ;
PROC    TREE ;
TITLE1 'Cluster Analysis of 28 Pupils' ;
TITLE2 'with regard to Ability Test Scores' ;
DATA    TREE ;
SET     TREE ;
        IF _NCL_ <=10 ;
PROC    PLOT DATA=TREE ;
        PLOT    _CCC_ * _NCL_ = _NCL_ ;
```

This program bases the cluster analysis on the squared Euclidean distance matrix $\mathbf{D}^2 = \{d_{ij}^2\}$ with

$$d_{ij}^2 = \sum_{l=1}^{p} (X_{il} - X_{jl})^2.$$

The observations were standardized prior to calculating the distances. Although an average linkage analysis was used in this case, a centroid linkage analysis or a Ward-analysis is also possible with this program.

The SAS PROC CLUSTER program does not produce a graphical representation of clusters but should precede the SAS PROC TREE program which produces a cluster map. The output below shows the cluster map of the 28 pupils.

Output:

```
                          Cluster Analysis of 28 Pupils
                          with regard to Ability Test Scores

                             NAME OF OBSERVATION OR CLUSTER
```

If for example, the pupils are to be divided into five fairly homogeneous groups with regard to achievement in the 18 ability tests the following grouping can be obtained from the cluster map.

$$(1, 12, 2, 7, 10, 21, 22, 27, 23, 25, 6) \qquad \text{cluster 1}$$
$$(5) \qquad \text{cluster 2}$$
$$(3, 8, 9, 28, 4, 13, 14, 19, 20, 18, 17, 26) \qquad \text{cluster 3}$$
$$(15, 24) \qquad \text{cluster 4}$$
$$(11, 16) \qquad \text{cluster 5}$$

The cubic clustering criterion (CCC) is standard output in the case of the SAS PROC CLUSTER program. Criterion values are given for $g = 1, 2, \ldots, k$ clusters where k usually is the largest integer smaller than $n/10$. The value of g associated with the largest CCC-value is the best choice for the number of clusters. A maximum CCC-value of larger than 2 indicates a meaningful cluster analysis. Values of CCC between 0 and 2 should be interpreted carefully, whereas negative CCC-values, having a descending trend for increasing g, give an indication that the underlying distribution is unimodal. Use of the cubic clustering criterion is recommended only for large data sets where each cluster contains at least 10 observation points. The CCC can therefore not be used meaningfully for determining an optimal value for g in the case of the 28 pupils' ability test scores.

Consider next, the BMDP-2M program.

Input: BMDP-2M program for cluster analysis

```
/PROBLEM     TITLE='Cluster Analysis of Ability Scores of 28 Pupils'.
/INPUT       VARIABLES=18.
             FORMAT='(18F3.0)'.
/VARIABLE    NAMES ARE X1,X2,X3,X4,X5,X6,X7,X8,X9,X10,X11,X12,
             X13,X14,X15,X16,X17,X18.
/PROCEDURE   LINK=SING.
             SUMOFSQ.
             SUMOFP=2.
             STAND.
/PRINT       DATA.
             DIST.
             VERT.
             SHADE.
/END
 19 21 21 18 20 21 15 14 15 13 15 16 19 19 19 20 17 17
 21 20 15 24 22 18 11 18 16 19 14 17 21 15 17 18 18 19
 18 19 16 18 18 23 11 13 13 15 11 11 15 18 13 15 18 13
              .
              .
              .
 22 17 21 17 17 22 10 14 16 16 13 08 13 18 21 12 13 15
 21 18 20 23 21 22 08 15 09 17 11 13 13 20 20 21 15 20
 21 22 19 20 18 17 11 15 12 14 11 10 11 13 14 14 15 14
/*
//
```

The BMDP-2M program uses either the single linkage or centroid linkage technique. It provides for most of the distance metrics discussed in Section 3. In this application the single linkage criterion based on the Euclidean distances of the standardized scores was used (Pearson distances).

Output:

(i) *Dendrogram of 28 pupils with regard to their ability test scores*

```
          BMDP2M  Cluster  Analysis  of  Ability  Scores  of  28  Pupils


    C  N
    0        1         2 2 2 2 1 2 1    2 1         2 1 2 1 1 2 1 1    1
    A  .   1 2 7 2 3 5 1 2 0 6 5 9 8 3 8 3 4 4 7 7 8 4 0 9 1 5 6 6

    S  L
       A
    E  B
       E
       L


AMALG.
DISTANCE
             *  *  *  *  *  *  *  *  *  *  *  *  *  *  *  *  *  *  *  *  *  *  *  *  *  *  *  *
    2.508    -+-  I  I  I  I  I  I  I  I  I  I  I  I  I  I  I  I  I  I  I  I  I  I  I  I  I
    2.913     I   I  I  I  I  I  I  I  I  -+-  I  I  I  I  I  I  I  I  I  I  I  I  I  I  I
    3.227     I   I  I  I  I  I  I  I  I  I   I  -+-  I  I  I  I  I  I  I  I  I  I  I  I
    3.271     I  -+-  I  I  I  I  I  I  I   I  I  I   I  I  I  I  I  I  I  I  I  I  I  I
    3.271     I   I   I  I  I  I  I  I  I   -+---  I   I  I  I  I  I  I  I  I  I  I  I  I
    3.423    --+--  I  I  I  I  I  I  I   I    I  I  I .I  I  I  I  I  I  I  I  I  I  I
    3.433     I    -+-  I  I  I  I  I    I      I  I  I  I  I  I  I  I  I  I  I  I  I  I
    3.483     I     I   I  I  I  I  I    --+---     I  I  I  I  I  I  I  I  I  I  I  I
    3.501     I     I   I  I  I  I  -----+-        I  I  I  I  I  I  I  I  I  I  I  I
    3.536     I     I   I  I  I  I  ------+-       I  I  I  I  I  I  I  I  I  I  I  I
    3.559     I     I  -+-  I           I          I  I  I  I  I  I  I  I  I  I  I  I
    3.562     I     I    -+--            I          I  I  I  I  I  I  I  I  I  I  I  I
    3.588     I    ---+---               I          I  I  I  I  I  I  I  I  I  I  I  I
    3.597     I      I                   I          I  I  I  I  I  -+-  I  I  I  I
    3.632    -----+----                  I          I  I  I  I  I  I   I   I  I  I  I
    3.644    -------+---------            I  I  I  I  I   I   I  I  I  I
    3.772              I                      I  I  I  I  --+-  I  I  I  I
    3.857                  -+----------------- I  I  I   I     I  I  I  I
    3.928                   -+----------------- I  I  I   I     I  I  I  I
    3.942              I                         I  I  I  -+---  I  I  I
    4.022                  -+----------------- I  I   I     I  I  I
    4.068              I                          I  ----+-    I  I  I
    4.086                  -+-----------------       I       I  I  I
    4.107             -----+-----------------        I  I  I
    4.324                  -+------------------------------ I  I
    4.395                  -+------------------------------- I
    4.476                  -+------------------------------
```

If a threshold value of 4.0 is considered eight clusters can be formed namely:

(1, 12, 7, 2, 23, 25, 21, 22, 10, 26, 15, 9, 28, 13, 8, 3, 4, 24); (14, 20, 19, 11); (17); (27);

(18); (5); (16); (6).

Comparing the above eight clusters with the five formed with the cluster map, shows some correspondence as well as some differences in the groupings. This is due to the fact that different clustering methods as well as different distance measures were used in the two applications.

(ii) *Shaded representation* (SHADE)

This representation portrays the ordered distance matrix in such a way that darkly shaded sections in the matrix point to great similarity and lightly shaded sections to little similarity between cases (pupils). The order in which the cases are arranged vertically and horizontally in the representation corresponds to the order in the dendrogram. The table below indicates the distances associated with the different shading symbols used in the representation.

```
DISTANCES  BETWEEN  CASES  REPRESENTED  IN  SHADED  FORM.
HEAVY  SHADING  INDICATES  SMALL  DISTANCES.


CASE    CASE
NO.     LABEL

   1            ▪
  12            ▪▪
   7            ▪▪▪
   2            ▪×▪▪
  23            ++▨▨▪
  25            ▪▨▪▪▪▪
  21            ▨▪▪▪▪▪▪
  22            ▪▪▪▪×▪▪▪
  10            ▪▪▪××××▪▪▪
  26            ××××+-+▨▪▪▪
  15            ×××+.---+×▪
   9            ++×++-++×××▪▪
  28            ×××××+××××▪▨▪▪
  13            ---.-.-.-×××▪▪
   8            ×××+++▨××▪×▪▪▪▪
   3            ▪▪▪×+×▪▪▪▪▪▪▪▪▪
   4            ...      ..-+▪▪▪+××
  24            ▪▪▪×+×++--▪××▨--×.▪
  17            ---...-+-▪××××××-+▪
  27            ▨×▨××▨▪▪▪×+××-××.--▪
  18            ×++-...++×▪▪▪×××▨××▨.▪
  14            ×+-.....--+▨×××+▨×-×-▨▪
  20            +-.....--+××▪▪×▨×-+-▪▪▪
  19            .             -×××▨+×▪.×  ▪▪▪▪
  11            .           .-▨×××-××.-  ×++▪▪
   5            ▪▪××.××+.    .   .- -     ▪
  16                        --.× .× .  -.-▨×  ▪
   6            ▨××▨×××▨++++++.++  ×-×.-.    .  ▪
```

```
THE  DISTANCES  HAVE  BEEN  REPRESENTED  ABOVE  IN  SHADED
FORM  ACCORDING  TO  THE  FOLLOWING  SCHEME
```

			LESS THAN		4.117
▪			LESS THAN		4.117
▨	FROM	4.117	TO		4.575
×	FROM	4.575	TO		5.116
X	FROM	5.116	TO		5.448
+	FROM	5.448	TO		5.989
-	FROM	5.989	TO		6.655
.	FROM	6.655	TO		7.653
			GREATER THAN		7.653

The following conclusions are, for example, evident from the above shaded representation. Pupil 6 is somewhat isolated and only shows little similarity with pupils 1, 12, 7, 2, 23 and 25. Pupil 3 shows a large degree of similarity to most of the pupils.

6. Digraphs

Various everyday phenomena as well as sophisticated systems can often be characterized by a twofold relationship. A systematical representation of a set of direction dependent twofold relationships is known as a digraph. A number of well known applications of the digraph are:

○ Representations of social relationships.
○ Flow diagrams in computer science and electronics.
○ Road, rail and telecommunication networks.
○ Representing and scheduling sporting events.
○ Tree diagrams in the theory of probability and game theory.

A digraph consists of a number of points (nodes) linked together by curves or lines also called edges. Each curve indicates the direction of the relationship between two points. Figure 5.4 shows a digraph of the preferences of 92 students on comparing five kinds of meat pairwise [Robinson and Foulds (1980)]. In this diagram an arrow pointing from pork to tongue indicates that the majority of the students prefer pork to tongue. Steak is obviously the favorite whereas tongue is not exactly popular.

The number of arrows ending at a specific node is called the outdegree of

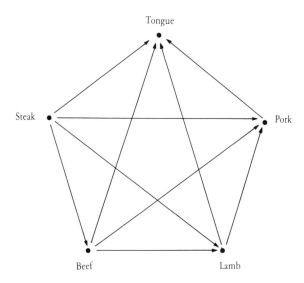

Figure 5.4. Digraph of 92 students' preferences in respect of five kinds of meat.

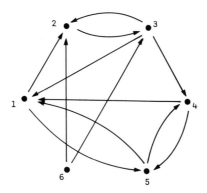

Figure 5.5. Friendship between six persons.

the node and the number of arrows leading from it, the indegree. Consequently beef has an indegree of 3 and an outdegree of 1.

As another example consider the digraph in Figure 5.5 which illustrates the social relationship between six persons.

The information in the above figure can also be presented in a so-called adjacency matrix, namely

	1	2	3	4	5	6	Outdegree (row total)
$\mathbf{X} = 1$	0	1	0	0	1	0	2
2	0	0	1	0	0	0	1
3	1	1	0	1	0	0	3
4	1	0	0	0	1	0	2
5	1	0	0	1	0	0	2
6	0	1	1	0	0	0	2
Indegree (column total)	3	3	2	2	2	0	12

where

$$X_{ij} = \begin{cases} 1 \text{ if } i \text{ considers } j \text{ as a friend} \\ 0 \text{ otherwise.} \end{cases}$$

In analyzing adjacency matrices the number of mutual friendships M, is vital. The value of M, together with the empirical distributions of the in and outdegrees paves the way for interesting conclusions concerning the type of relationship existing between persons (for instance clique formation or leader-follower).

In order to show how cluster analysis can be used for detecting friendship groups, consider an adjacency matrix representing the friendships between 18 monks [Sampson (1969)]. The adjacency matrix is given below in Table 5.3.

The M value (15 in this case) as well as the in and outdegrees point to clique

Table 5.3. Adjacency matrix for friendships between 18 monks

X =	1	2	3	4	5	6	7	8	9	10	11	12	13	14	15	16	17	18	Outdegree
1	0	1	1	0	1	0	0	0	0	0	0	0	0	0	1	0	0	0	4
2	0	0	1	0	1	1	0	0	0	0	0	0	0	0	0	0	0	0	3
3	0	1	0	0	0	0	1	1	0	0	0	0	0	0	0	0	0	0	3
4	0	1	1	0	1	0	0	0	0	0	0	0	0	0	0	0	0	0	3
5	0	1	0	1	0	1	0	0	0	0	0	0	0	0	0	0	0	0	3
6	0	1	0	0	1	0	1	0	0	0	0	0	0	0	0	0	0	0	3
7	0	0	1	1	0	0	0	1	0	0	0	0	0	0	0	0	0	0	3
8	0	0	0	0	0	0	0	1	1	1	0	0	0	0	0	0	0	0	3
9	0	0	0	0	0	0	0	1	0	1	1	0	0	0	0	0	0	0	3
10	0	0	0	0	0	0	0	1	1	0	0	0	1	0	0	0	0	0	3
11	0	0	0	0	0	0	0	0	1	0	0	1	1	0	0	0	0	0	3
12	0	0	0	0	0	0	0	0	0	1	0	0	1	1	0	0	0	0	3
13	0	0	0	0	0	0	0	1	0	1	0	1	0	0	0	0	0	0	3
14	0	0	0	0	0	0	0	0	0	1	0	0	1	0	0	1	0	0	3
15	0	0	0	0	0	0	0	0	1	0	0	0	1	0	0	0	0	1	3
16	0	0	0	0	0	0	0	0	0	1	0	0	0	0	1	0	1	1	4
17	0	1	0	0	0	0	0	0	0	0	0	0	0	0	0	1	0	1	3
18	0	0	0	0	0	0	0	0	0	0	1	0	0	0	0	1	1	0	3
Indegree	0	6	4	2	4	2	2	6	4	6	2	2	5	1	2	3	2	3	56

Table 5.4. Similarity matrix for friendships between 18 monks

C =	1	2	3	4	5	6	7	8	9	10	11	12	13	14	15	16	17
1																	
2	0.83																
3	0.72	0.67															
4	0.94	0.89	0.78														
5	0.72	0.78	0.78	0.78													
6	0.72	0.67	0.89	0.78	0.78												
7	0.83	0.89	0.67	0.89	0.78	0.67											
8	0.61	0.67	0.67	0.67	0.67	0.67	0.67										
9	0.61	0.67	0.78	0.67	0.67	0.78	0.67	0.67									
10	0.61	0.67	0.78	0.67	0.67	0.89	0.67	0.89	0.78								
11	0.61	0.67	0.78	0.67	0.67	0.78	0.67	0.78	0.83	0.89							
12	0.61	0.67	0.78	0.67	0.67	0.89	0.67	0.89	0.78	0.89	0.78						
13	0.61	0.67	0.78	0.67	0.67	0.78	0.67	0.78	0.78	0.78	0.78	0.89					
14	0.61	0.67	0.67	0.67	0.67	0.89	0.67	0.89	0.67	0.78	0.78	0.89	0.78				
15	0.72	0.67	0.78	0.78	0.78	0.78	0.67	0.78	0.67	0.78	0.67	0.78	0.67	0.78			
16	0.67	0.61	0.61	0.61	0.61	0.72	0.61	0.72	0.61	0.72	0.72	0.61	0.61	0.61	0.72		
17	0.61	0.67	0.67	0.67	0.67	0.78	0.67	0.78	0.78	0.67	0.67	0.78	0.78	0.78	0.78	0.72	
18	0.61	0.67	0.67	0.67	0.67	0.78	0.67	0.78	0.78	0.67	0.67	0.78	0.78	0.78	0.67	0.72	0.89

formation. In an attempt to identify the various cliques, the similarity matrix $\mathbf{C}$ of Sokal and Michener, as defined in Section 3, was calculated, and a single linkage cluster analysis carried out. The matrix $\mathbf{C}$ appears in Table 5.4 and the dendrogram is given in Figure 5.6.

As an illustration of obtaining the elements of $\mathbf{C}$ consider, for example, c_{21}. The value c_{21} indicates the "closeness" between the first two monks and is obtained from the first two rows of the adjacency matrix $\mathbf{X}$, by calculating

$a = $ number of columns with $X_{1j} = X_{2j} = 1$

$d = $ number of columns with $X_{1j} = X_{2j} = 0, \qquad j = 1, 2, \ldots, 18,$

and substituting these values in the definition of c_{21}. From this it follows that

$$c_{21} = (2 + 13)/18$$

$$= 0.83.$$

From Figure 5.6 it follows that close friendship groups exist namely: $(1, 4, 2, 7)$; $(3, 6)$; $(8, 10, 12, 14, 11, 13)$ and $(17, 18)$. The other monks, especially number 16, do not seem to have such close ties with each other.

7. Spanning Trees

Consider a symmetrical digraph (i.e., a digraph in which all relationships are mutual). In this case direction is no longer of any consequence and all nodes are linked merely by straight lines. Now remove certain of the lines between the nodes in a random way. When no further lines can be removed without "dislocating" the diagram, a spanning tree is obtained. A weight is normally associated with every line (branch) of the spanning tree and a minimal spanning tree (MST) is obtained by minimizing the sum of the line weights. Although the line weight is usually determined by the distance between the two respective points, it can for instance, also represent cost, time, etc. depending on the problem under consideration. The minimal spanning tree will now be determined by means of a method proposed by Kruskal (1956).

Figure 5.7 shows all the routes that link seven cities to one another. The distances (km) between the cities are given at each line. Suppose the seven cities should be linked with six freeways by means of the shortest possible route.

Kruskal's method amounts to the following: Consider all the route distances in an ascending order. The shortest distance namely CG forms the first branch of the tree. Each route is now considered and drawn as a further branch of the tree unless it completes a cycle with previously formed branches. Figure 5.8 gives the minimal spanning tree obtained in this way. The ordered distances are as follows:

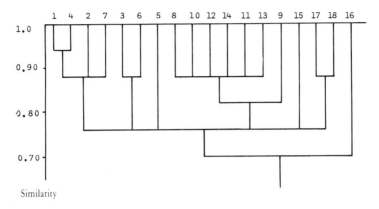

Figure 5.6. Dendrogram of friendships between 18 monks.

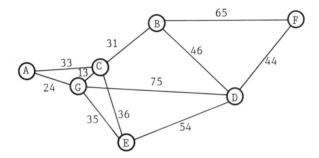

Figure 5.7. Route distances between seven cities.

GC, AG, CB, AC, GE, CE, DF, BD, ED, BF, GD.

The order of construction of the minimal spanning tree is as follows:

GC, AG, CB, GE, DF, BD.

Note for instance, that after constructing (*GC, AG, CB*) the shortest remaining distance is *AC*. Since a cycle *AGC* results if *AC* is contructed (which is contrary to the definition of a spanning tree) this line is omitted.

A single linkage cluster analysis can be obtained from a minimal spanning tree by successively eliminating the branches from the longest to the shortest. With regard to the above example, two groups (clusters) of cities can be obtained, namely (*A, G, E, C, B*); (*D, F*) if the longest branch *BD* is removed. Three groups namely (*A, G, E, C, B*); *D*; *F* can be obtained if the second longest branch *DF* is removed, etc. For further applications of minimal spanning trees, see Chapters 6 and 11.

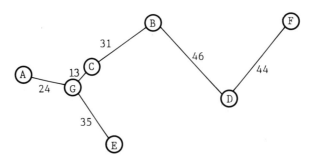

Figure 5.8. Minimal spanning tree for freeways between seven cities.

8. Cluster Analysis of Variables

Frequently it is important to divide the p variables (rather than the n cases) into fairly homogeneous groups or clusters on the basis of the data matrix $\mathbf{X}$. If, for example, the tests in a test battery can be broken up into fairly homogeneous clusters and a single test (variable) is chosen from each such cluster, a simplified test battery can be compiled which will provide roughly the same amount of information on the test persons as the original battery.

A matrix of Pearson correlations between the p variables is the obvious starting point for a cluster analysis of quantitative variables. The SAS PROC VARCLUS program uses a cluster analysis of Harman (1976) which amounts to choosing those clusterings of the variables that will maximize the total sum of the variation explained by the first principal components (see Chapter 6) of the various clusters of variables. Using the SAS program PROC TREE as illustrated in Section 5 in conjunction with the SAS PROC VARCLUS program, a cluster map of the variables can be obtained.

Hierarchical cluster analysis, through any of the techniques discussed in Section 4, can be applied to a similarity matrix or dissimilarity (distance) matrix obtainable from $\mathbf{R}$, the matrix of Pearson correlations. Although $\mathbf{R}$ can be used as the similarity matrix, $\mathbf{C} = \{|r_{ij}|\}$ is usually used in the cluster analysis. Alternatively, $\mathbf{R}$ can be transformed into a dissimilarity matrix $\mathbf{D}$ by considering $\mathbf{D} = \{1 - r_{ij}^2\}$ or $\mathbf{D} = \{1 - |r_{ij}|\}$. A dendrogram (see Figure 5.9) or cluster map obtained from a hierarchical analysis can now be used in the same way as illustrated in Sections 4 and 5 with a view to identifying homogeneous clusters of variables.

An alternative graphical procedure that may be used in the cluster analysis of variables is a profile chart of variables. In this graphical representation the correlation profile of each variable is plotted (see Figure 5.10).

For illustrative purposes we consider the data in Table 4.1 and carry out a cluster analysis on the six variables, X_1 to X_6. Table 5.5 shows the distance matrix $\mathbf{D} = \{1 - r_{ij}^2\}$ obtained for this data set.

Table 5.5. Dissimilarity matrix of X_1 to X_6 based on five pupils' scores

Ability test	X_1	X_2	X_3	X_4	X_5	X_6
X_1	0					
X_2	0.646	0				
X_3	0.833	0.222	0			
X_4	0.930	0.664	0.708	0		
X_5	0.991	0.396	0.195	0.924	0	
X_6	0.995	0.387	0.172	0.559	0.211	0

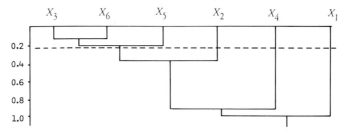

Distance between clusters

Figure 5.9. Dendrogram of six ability tests.

A complete linkage cluster analysis produced the dendrogram in Figure 5.9 whereas Figure 5.10 shows the profile chart for variables X_1 to X_6. A meaningful threshold value for identifying clusters from Figure 5.9 is $d_0^2 = (1 - r_0^2) = 1 - (0.8783)^2 = 0.228$, where r_0 is the $2\frac{1}{2}\%$ upper percentage point of the sample distribution of r, with $n = 5$ under the hypothesis that there is no correlation between two variables. Values of r_0, for different values of n, are found in the majority of statistical tables. From Figure 5.9 this threshold value suggests the following clusters:

$$(X_3, X_6, X_5); (X_2); (X_4); (X_1).$$

From Figure 5.10 it appears that X_1 and X_4 differ considerably from the other four variables.

The BMDP-1M program was used next for carrying out a cluster analysis of the 18 ability tests (X_1 to X_{18}) in respect of the 28 pupils (see Table 1.1). This program is capable of carrying out single linkage, complete linkage or average linkage analyses, using the matrix $\mathbf{C} = \{|r_{ij}|\}$ as the similarity matrix. Other similarity measures however, all based on the matrix $\mathbf{R}$, are also available. The complete linkage method was used in the application below.

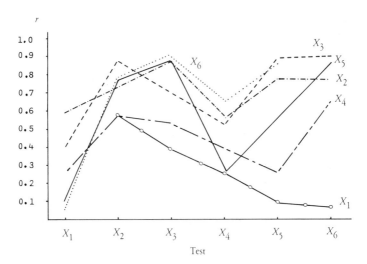

Figure 5.10. Profile chart of six ability tests.

Input: BMDP-1M program for compiling a correlation tree diagram

```
/PROBLEM      TITLE='Cluster Analysis of Ability Scores of 28 Pupils'.
/INPUT        VARIABLES=18.
              FORMAT='(18F3.0)'.
/VARIABLE     NAMES ARE X1,X2,X3,X4,X5,X6,X7,X8,X9,X10,X11,X12,
              X13,X14,X15,X16,X17,X18.
/PROCEDURE    LINK=COMP.
              VAR=X1,X2,X3,X4,X5,X6,X7,X8,X9,X10,X11,X12,X13,X14,
              X15,X16,X17,X18.
/PRINT        CORR.
              SHADE.
              TREE.
              EXPL.
/END
 19 21 21 18 20 21 15 14 15 13 15 16 19 19 19 20 17 17
 21 20 15 24 22 18 11 18 16 19 14 17 21 15 17 18 18 19
 18 19 16 18 18 23 11 13 13 15 11 11 15 18 13 15 18 13
              .
              .
              .
 22 17 21 17 17 22 10 14 16 16 13 08 13 18 21 12 13 15
 21 18 20 23 21 22 08 15 09 17 11 13 13 20 20 21 15 20
 21 22 19 20 18 17 11 15 12 14 11 10 11 13 14 14 15 14
/*
//
```

Output:

(i) *Correlation tree diagram for cluster analysis of variables*

The output below shows a tree diagram superimposed on the matrix of absolute correlations (multiplied by 100). The variables X_1 to X_{18} have been arranged vertically and horizontally in cluster order (in accordance with the dendrogram). The dotted lines of the tree diagram define possible clusters of variables. The similarity between the variables in each cluster can be read directly from the entries in the matrix.

```
TREE PRINTED OVER ABSOLUTE CORRELATION MATRIX.
CLUSTERING BY MAXIMUM DISTANCE METHOD.
     VARIABLE
NAME       NO.
           -------------------------------------------------------------/
X1       (   1)  61 67/48/42 40 40 46 40/41 55 29 19 42 44 48/15 29/
                      /  /                    /                 /   /
                   ----/  /                /                 /      /
X6       (   6)  77/53/53 45 58 60 36/46 41 48 27 58 49 53/ 9 17/
                      /  /            /                 /    /
                    /  /            /                 /    /
X10      (  10)/66/51 41 45 51 39/57 50 53 20 28 41 35/ 2  8/
                    /  /          /                 /    /
                  /  /          /                 /    /
X4       (   4)/47 47 56 59 41/50 58 46 18 42 42 46/ 2 19/
                    /                        /    /
                 --------------/                /    /
X5       (   5)  67/65 75/59/37 42 55 13 42 28 51/41 42/
                    /    /  /                 /    /
                  /    /  /  /              /    /
X12      (  12)/62 64/66/42 60 55  8 38 37 36/31 31/
                    /  /          /              /
                 ----/  /                /         /
X11      (  11)  79/55/35 44 46 10 47 41 62/29 47/
                    /  /                 /    /
                  /  /                 /    /
X17      (  17)/46/46 60 56 27 49 53 55/35 44/
                  /                 /    /
                /                 /    /
X18      (  18)/53 52 44 28 55 37 50/ 1 37/
                                 /    /
               --------------------/    /
X2       (   2)  64/61/45/45 55 28/14 16/
                    /  /  /        /    /
                  /  /  /  /     /    /
X8       (   8)/56/51/51 61 46/21 33/
                    /  /        /    /
                  /  /        /    /
X16      (  16)/34/29 32 38/20 26/
                  /        /    /
                /        /    /
X14      (  14)/34 55 56/22 25/
                      /    /
                 -------/    /
X3       (   3)  62/59/ 2 38/
                    /  /    /
                  /  /    /
X9       (   9)/60/17 39/
                  /    /
                /    /
X15      (  15)/23 32/
                    /
                 ----/
X7       (   7) 50/
                /
              /
X13      (  13)/
```

From this output a possible division into clusters is:

(X_1, X_6, X_{10}, X_4) with $|r_{ij}| \geq 0.48$ (Classification)
$(X_5, X_{12}, X_{11}, X_{17}, X_{18})$ with $|r_{ij}| \geq 0.46$ (Verbal Reasoning, Word Analogies)
(X_2, X_8, X_{16}) with $|r_{ij}| \geq 0.56$ (Figure Analogies)
(X_{14})

(X_3, X_9, X_{15}) with $|r_{ij}| \geqq 0.59$ (Pattern Completion)
(X_7, X_{13}) with $|r_{ij}| = 0.50$ (Number Series)

The names in brackets above indicate the tests which are largely represented by the variables in the different clusters.

(ii) *Shaded representation* (SHADE)

In this shaded representation darkly shaded sections indicate large absolute correlations whereas lightly shaded sections indicate little similarity between variables. The table below explains the shading symbols.

```
ABSOLUTE VALUES OF CORRELATIONS IN SORTED AND SHADED FORM

-------------------------------------------------------------

 1 X1        ■
 6 X6        ⊠■
10 X10       ⊠■■
 4 X4        X⋈⊠■
 5 X5        X⋈⋈X■
12 X12       +XXX⊠■
11 X11       X⋈X⋈⊠⊠■
17 X17       X⊠⋈⋈■⊠■■
18 X18       X++X⋈⊠⋈X■
 2 X2        XX⋈⋈+X+X⋈■
 8 X8        ⋈X⋈⋈X⊠X⊠⋈⊠■
16 X16       -X⋈X⋈⋈X⋈X⊠⋈■
14 X14       .-...  --X⋈+■
 3 X3        X⋈-XX+XX⋈X⋈-+■
 9 X9        XXXX-+X⋈+⋈⊠+⋈⊠■
15 X15       X⋈+X⋈+⊠⋈⋈-X+⋈⋈⊠■
 7 X7        .   X+-+ .-.- .-■
13 X13       -. .X+XX+.+--+++⋈■


THE ABSOLUTE VALUES OF
THE MATRIX ENTRIES HAVE BEEN PRINTED ABOVE IN SHADED FORM
ACCORDING TO THE FOLLOWING SCHEME
                  LESS THAN OR EQUAL TO           0.115
     .      0.115  TO AND INCLUDING               0.211
     -      0.211  TO AND INCLUDING               0.308
     +      0.308  TO AND INCLUDING               0.405
     X      0.405  TO AND INCLUDING               0.501
     ⋈      0.501  TO AND INCLUDING               0.598
     ⊠      0.598  TO AND INCLUDING               0.695
     ■                    GREATER THAN            0.695
```

The shaded representation inter alia shows that X_{14} and to some extent (X_7, X_{13}) also, are rather dissimilar to the other variables.

9. Application of Cluster Analysis to Fitness/ Cholesterol Data

As a final application a cluster analysis of cases as well as of variables was done on the fitness/cholesterol data set (see Chapter 1). Groups 1, (weight-lifters), 2 (control) and 3 (marathon athletes) were considered with respect to the fitness variables $X_2, X_3, \ldots, X_{15}$.

Clustering of Cases

Input: SAS PROC CLUSTER/TREE program using Ward's method based on Pearson distances

```
DATA    FITNESS ;
INPUT   NUMBER X1-X15 ;
CARDS ;
  1 1   22 179.2 107.1 15.2 3.0 8.8 1.0   92 130 19   15   24   24   183.6
  2 1   30 183.0 112.2 20.3 4.6 9.9 1.0   92 114 18   19   22   23   195.6
  3 1   26 175.7  78.0 17.5 3.7 6.6 1.7   85  62 18   22   23   24   106.4
          .
          .
          .
 55 3   27 174.9  74.1 17.4 3.6 5.3 2.1   68  46 18   18   24   24   101.1
 56 3   23 171.2  69.8 14.3 2.9 6.3 2.1   80  61 21   19   27   29   104.8
 57 3   24 176.3  72.4 15.3 3.1 5.5 2.6   73  81 20   18   26   28   113.3
;
PROC    CLUSTER STD METHOD=W OUTTREE=TREE ;
ID      NUMBER ;
VAR     X2-X15 ;
PROC    TREE ;
TITLE1  'Cluster Analysis of Groups 1 to 3' ;
TITLE2  'with regard to Fitness Variables' ;
DATA    TREE ;
SET     TREE ;
        IF _NCL_ <=10 ;
PROC    PLOT DATA=TREE ;
        PLOT   _CCC_*_NCL_=_NCL_  / VREF=0 HREF=0
                                    VAXIS = -4 TO 0 BY  1.0
                                    HAXIS =0 TO 10 BY 1
                                    VPOS=25 HPOS=40 ;
```

Output:

Since 4 of the 57 cases contain missing observations in one or more of the fitness variables only 53 cases were analyzed.

(i) *Cluster map of cases*

Cluster Analysis of Groups 1 to 3
with regard to Fitness Variables

NAME OF OBSERVATION OR CLUSTER

```
                3 1   2 3 5 4 2 5 5 1 2 4 2 2 3 1 3 5 2 2 2 4 5 4   1 1     1 1 1 2   1   3 5 4 3 4 4 3 3 3 5 3 4 5
                1 2 9 8 5 3 4 5 5 7 3 6 7 7 6 1 0 8 3 8 7 3 2 7 9 5 2 8 5 2 0 7 8 1 3 9 5 6 4 4 6 0 6 2 2 3 0 4 1 4 9 4 1
N      1 +XXXXXXXXXXXXXXXXXXXXXXXXXXXXXXXXXXXXXXXXXXXXXXXXXXXXXXXXXXXXXXXXXXXXXXXXXXXXXXXXXXXXXXXXXXXXXXX
U      2 +XXXXXXXXXXXXXXXXXXXXXXXXXXXXXXXXXXXXXXXXXXXXXXXXXXXXXXXXXXXXXXXXXXXXXXXXXXXXXXXXXXXXXXXXXXXXXXX
M      3 +XXXXXXXXX XXXXXXXXXXXXXXXXXXXXXXXXXXXXXXXXXXXXXXXXXXXXXXXXXX XXXXXXXXXXXXXXXXXXXXXXXXXXXXXXXXXX
B      4 +XXXXXXXXX XXXXXXXXXXXXXXXXXXXXXXXXXXXXXXXXXXXXXXXXXXXXXXXXXX XXXXXXXXXXXXXXXXXXXXXXXXXXXXXXXXXX
E      5 +XXXXXXXXX XXXXXXXXXXXXXXXXXXXXXXXXXXXXXXXXXXXXXX XXXXXXXXXXX XXXXXXXXXXXXXXXXXXXXXXXXXXXXXXXXXX
R      6 +XXX XXXXX XXXXXXXXXXXXXXXXXXXXXXXXXXXXXXXXXXXXXX XXXXXXXXX  XXXXXXXXXXXXXXXXXXXXXXXXXXXXXXXXXX
       7 +XXX XXXXX XXXXXXXXXXXXXXXXXXXXXXXXXXXXXXXXX XXXXXXX XXXXXXX  XXXXXXXXXXXXXXXXXXXXXXXXXXXXXXXXX
O      8 +XXX XXXXX XXXXXXXXXXXXXXXXXXXXXXXXXX XXXXXXX XXXXXXX XXXXXXX  XXXXXXXXXXXXXXXXXXXXXXXXXXX XXXXX
F      9 +XXX XXXXX XXXXXXXXXXXXXXXXXXXXXXXXXX XXXXXXX XXXXXXX XXXXXXXXXXXXXXXXX  XXXXXXXXXXXXXXXXXX XXXXX
      10 +XXX XXXXX XXXXXXXXXXXXXXXXXXXXXXXXXX XXXXXXX XXXXXXX XXXXXXXXXXXXXX XXX XXX XXXXXXXXXXXXXXXXX XXXXX
C     11 +XXX XXXXX XXXXXXXXXXXXXXXXXXXXXXXXXX XXXXXXX XXXXXXX XXXXXXX XXX XXX XXX XXXXXXXXXX XXXXXX XXXXX
L     12 +XXX XXX . XXXXXXXXXXXXXXXXXXXXXXXXXX XXXXXXX XXXXXXX XXXXXXX XXX XXX XXXXXXXXXXX XXXXXX XXXXX
U     13 +XXX XXX . XXXXXXXXXXXXXXXX XXXXXXXXX XXXXXXX XXXXXXX XXXXXXX XXX XXX XXXXXXXXXXX XXXXXX XXXXX
S     14 +XXX XXX . XXXXXXXXXXXXXXXX XXXXXXXXX XXXXXXX XXXXXXX XXXXXXX XXX XXX XXXXXXX XXXXX XXXXXXX XXXXX
T     15 +XXX XXX . XXXXXXXXXXXXXXXX XXXXXXXXX XXXXXXX XXXXXXX XXXXXXX XXX XXX XXXXXXX XXXXX XXXXXXX XXXXX
E     16 +XXX XXX . XXXXXXXXXX XXXXX XXXXXXXXX XXXXXXX XXXXXXX XXXXXXX XXX XXX XXXXXXX XXXXX XXXXXXX . XXX
R     17 +XXX XXX . XXXXXXXXXX XXXXX XXXXXXXXX XXXXX . XXXXXXX XXXXXXX XXX XXX XXXXXXX XXXXX XXXXXXX . XXX
S     18 +XXX XXX . XXXXXXXXXX XXXXX XXXXXXXXX XXXXX . XXXXXXX XXXXXXX XXX XXX XXXXXXX XXXXX XXXXXXX . XXX
      19 +XXX XXX . XXXXXXXXXX XXXXX XXXXXXXXX XXXXX . XXX XXXXX XXXXXX XXX XXX XXXXXXX XXXXX XXXXXXX . XXX
      20 +XXX XXX . XXXXXXXXXX XXXXX XXXXX XXXXX XXXXX . XXX XXXXX XXXXXX XXX XXX XXXXXXX XXXXX XXXXXXX . XXX
      21 +XXX XXX . XXXXXXXXX XXXXX XXXXX XXXXX XXX . XXX XXXXX XXXXXXXX XXX XXX XXXXXX XXXXX XXXXXXX . XXX
      22 +XXX XXX . XXXXXXXXX XXXXX XXXXX XXXXX XXX . XXX XXXXX XXXXXXXX XXX XXX XXX XXXXXX XXXXX XXXXXXX . XXX
      23 +XXX XXX . XXXXXXXXX XXXXX XXXXX XXXXX XXX . XXX XXXXX XXXXXXXX XXX XXX . XXXXXX XXXXX XXXXXXX . . .
      24 +. . XXX . XXXXXXXXX XXXXX XXXXX XXXXX XXX . XXX XXXXX XXXXXXXX XXX XXX . XXXXXX XXXXX XXXXXXX . . .
      25 +. . XXX . XXXXXXXXX XXXXX XXXXX XXXXX XXX . XXX XXXXX XXXXXXXX XXX XXX . . XXXXX XXXXX XXXXXXX . . .
      26 +. . XXX . XXXXXXXXX XXXXX XXXXX XXXXX XXX . XXX XXXXX XXXXXXXX XXX . . . XXXXX XXXXX XXXXXXX . . .
      27 +. . XXX . XXXXXXXXX XXXXX XXXXX XXXXX XXX . XXX XXXXX XXXXXXXX XXX . . . XXXXX XXXXX XXXXXXX . . .
      28 +. . XXX . XXXXXXXXX XXXXX XXXXX XXXXX XXX . XXX XXXXX XXXXX XXX XXX . . XXXXX XXXXX XXX XXX . . .
      29 +. . XXX . XXXXXXXXX XXXXX XXXXX XXX . XXXXX XXX . . XXXXX XXXXX XXX XXX . XXXXX XXXXX XXX XXX . . .
      30 +. . XXX . XXXXXXXXX XXXXX XXX . XXXXX XXX . . XXXXX XXX XXX . XXXXX XXXXX XXX XXX . . .
      31 +. . XXX . XXX XXXXX XXXXX XXX . XXXXX XXX . . . XXXXX XXXXX XXX XXX . XXXXX XXXXX XXX XXX . . .
      32 +. . XXX . XXX XXXXX XXXXX XXX . XXXXX XXX . . . XXXXX XXXXX . XXX . XXXXX XXXXX XXX XXX . . .
      33 +. . XXX . XXX XXXXX XXXXX XXX . XXXXX XXX . . . XXXXX XXX . . XXX . XXXXX XXXXX XXX XXX . . .
      34 +. . XXX . XXX XXXXX XXXXX XXX . XXXXX XXX . . . XXX . XXX . XXXXX . XXX XXX XXX . . .
      35 +. . . . XXX XXXXX XXXXX XXX . XXXXX XXX . . . XXX . XXX . XXXXX . XXX XXX XXX . . .
      36 +. . . . XXX XXXXX XXXXX XXX . XXXXX XXX . . . XXX . XXX . XXXXX . XXX XXX . . .
      37 +. . . . XXX XXXXX XXXXX XXX . XXX . XXX . . XXXXX XXX . . XXX . XXXXX . XXX XXX . . .
      38 +. . . . . XXXXX XXXXX XXX . XXX . XXX . . XXXXX XXX . . XXX . XXXXX . XXX XXX . . .
      39 +. . . . . XXXXX XXXXX XXX . XXX . XXX . . XXXXX XXX . . XXX . XXXXX . XXX . . .
      40 +. . . . . XXXXX XXXXX XXX . XXX . XXX . . . XXX . . XXX . XXXXX . XXX . . .
      41 +. . . . . XXX . XXXXX XXX . XXX . XXX . . . XXX . . XXX . XXXXX . XXX . . .
      42 +. . . . . XXX . XXXXX XXX . XXX . XXX . . . XXX . XXX . XXX . XXX . . .
      43 +. . . . . XXX . XXXXX XXX . XXX . XXX . . . . XXX . . XXX . XXX . . .
      44 +. . . . . XXX . XXX XXX . XXX . XXX . . . XXX . XXX . XXX . XXX . . .
      45 +. . . . . XXX . XXX XXX . XXX . XXX . . . . XXX . XXX . XXX . . .
      46 +. . . . . XXX . XXX XXX . XXX . . . . XXX . XXX . XXX . . .
      47 +. . . . . . . XXX XXX . XXX . . . . XXX . . XXX . XXX . . .
      48 +. . . . . . . XXX XXX . XXX . . . . . XXX . XXX . XXX . . .
      49 +. . . . . . . . XXX . XXX . . . . . XXX . XXX . XXX . . .
      50 +. . . . . . . . XXX . . . . . XXX . XXX . XXX . . .
      51 +. . . . . . . . XXX . . . . . . XXX . XXX . . .
      52 +. . . . . . . . . XXX . . . . . . . . . . .
      53 +. . . . . . . . . . . . . . . . . . . . .
```

(ii) *Cubic clustering criterion*

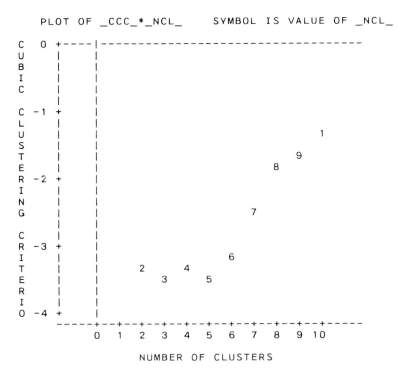

```
        PLOT  OF  _CCC_*_NCL_        SYMBOL  IS  VALUE  OF  _NCL_

C     0 +----|----------------------------------
U       |    |
B       |    |
I       |    |
C       |    |
        |    |
C    -1 +    |
L       |    |
U       |    |                                        1
S       |    |
T       |    |                                    9
E       |    |                               8
R    -2 +    |
I       |    |
N       |    |
G       |    |                          7
        |    |
C       |    |
R    -3 +    |
I       |    |                       6
T       |    |      2        4
E       |    |         3        5
R       |    |
I       |    |
O    -4 +    |
        ----+--+--+--+--+--+--+--+--+--+-----
            0   1   2   3   4   5   6   7   8   9  10

                    NUMBER  OF  CLUSTERS

NOTE:        1 OBS  HAD  MISSING  VALUES  OR  WERE  OUT  OF  RANGE
```

Although the CCC plot indicates that no natural clusters exist, three clusters were formed to ascertain the degree of correspondence between these three clusters and the original three groups. Using the cluster map the three clusters are:

Cluster 1:
Cases 1, 2, 9, 38, 15.

Cluster 2:
Cases 3, 24, 35, 55, 47, 23, 56, 57, 17, 26, 41, 20, 28, 33, 18, 37, 53, 22, 27, 29, 45, 52, 48, 5, 12, 10, 7, 8, 11, 13, 19, 25, 6, 14.

Cluster 3:
Cases 4, 36, 50, 46, 32, 42, 43, 30, 34, 31, 54, 39, 44, 51.

Comparing the three groups with these three clusters one can compile the following table.

Cluster	Groups		
	Weightlifters	Control	Marathon athletes
1	4	0	1
2	11	14	9
3	1	5	8

Clusters 1 and 3 largely consist of weightlifters and marathon athletes respectively. Most of the students are in Cluster 2.

Clustering of Variables

Input: SAS PROC VARCLUS/TREE program using Harman's method based on the correlation matrix.

```
DATA    FITNESS ;
INPUT   NUMBER X1-X15 ;
CARDS ;
  1 1  22 179.2 107.1 15.2 3.0 8.8 1.0  92 130 19  15  24  24  183.6
  2 1  30 183.0 112.2 20.3 4.6 9.9 1.0  92 114 18  19  22  23  195.6
  3 1  26 175.7  78.0 17.5 3.7 6.6 1.7  85  62 18  22  23  24  106.4
           .
           .
           .
 55 3  27 174.9  74.1 17.4 3.6 5.3 2.1  68  46 18  18  24  24  101.1
 56 3  23 171.2  69.8 14.3 2.9 6.3 2.1  80  61 21  19  27  29  104.8
 57 3  24 176.3  72.4 15.3 3.1 5.5 2.6  73  81 20  18  26  28  113.3
;
PROC    VARCLUS OUTTREE=TREE ;
VAR     X2-X15 ;
PROC    TREE ;
TITLE1 'Cluster Analysis of Fitness Variables' ;
TITLE2 'with regard to Groups 1 to 3' ;
```

Output:

The 53 cases with no missing observations were used and according to Harman's criterion the 14 variables can be grouped in at most five clusters. The resulting cluster map is given below.

Cluster map of variables

```
              Cluster Analysis of Fitness Variables
                   with regard to Groups 1 to 3

                     NAME OF VARIABLE OR CLUSTER

                                                          X     X      X      X
              X     X     X     X     1     1     X     X     X     1     1     1     1
              X     X     X     X     1     1     X     X     X     X     1     1     1     1
              5     6     2     3     1     2     4     7     8     9     0     5     3     4
     N    1  +XXXXXXXXXXXXXXXXXXXXXXXXXXXXXXXXXXXXXXXXXXXXXXXXXXXXXXXXXXXXXXXXXXXXXXXXXXXXXXX
     U    2  +XXXXXXXXXXXXXXXXXXXXXXXXXXXXXXXXXXXX     XXXXXXXXXXXXXXXXXXXXXXXXXXXXXXXXXXXXXXXXXXXX
     M    3  +XXXXXXXXXXXXXXXXXXXX      XXXXXXX     XXXXXXXXXXXXXXXXXXXXXXXXXXXXXXXXXXXXXXXXXX
     B    4  +XXXXXXXXXXXXXXXXXXXX      XXXXXXX     XXXXXXXXXXXXXXXXXXXXXXXXXXXXXXXXXXXXXX     XXXXXXX
     E    5  +XXXXXXX      XXXXXXX      XXXXXXX     XXXXXXXXXXXXXXXXXXXXXXXXXXXXXXXXXXX       XXXXXXX
     R    6  +.           .           .            .           .           .           .
          7  +.           .           .            .           .           .           .
     O    8  +.           .           .            .           .           .           .
     F    9  +.           .           .            .           .           .           .
         10  +.           .           .            .           .           .           .
     C   11  +.           .           .            .           .           .           .
     L   12  +.           .           .            .           .           .           .
     U   13  +.           .           .            .           .           .           .
     S   14  +.           .           .            .           .           .           .
```

It is interesting to note that Cluster 4 contains all the strength variables and Clusters 3 and 5 all the reaction time variables.

10. Other Graphical Techniques of Cluster Analysis

Kleiner and Hartigan (1981) have developed tree representations which simultaneously point out the similarity between variables and between observation vectors. Their procedure entails the following:

(a) Obtain a dendrogram of the p variables by carrying out any hierarchical cluster analysis on the data matrix $\mathbf{X}$.

(b) Use this dendrogram to draw a prototype tree depicting the correlation structure of the variables. If, for example, $p = 6$, the six leaves represent the six variables and the trunk and branches correspond to the branching within the dendrogram. The similarity between the variables is represented by the angles between the branches and leaves as well as the thickness of the trunk.

(c) Subsequently a tree is drawn for each observation vector (case) by making the length of the leaves proportional to the observations and choosing the length of the trunk and branches proportional to the arithmetic mean of the observed variables above the trunk or branches. Two typical trees are shown in the schematic presentation below.

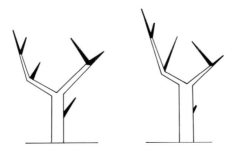

For n cases, n trees can be formed according to steps (a)–(c). The n trees formed in this manner all have the same basic form, but differ widely in respect of the overall size. On the basis of the forms of the trees, cases can then be divided into clusters. This procedure can be modified in various ways, for instance by obtaining castles, which entails making the angle between branches $0°$ and drawing the branches alongside each other [see Kleiner and Hartigan (1981)]. A computer program for representing trees and castles was developed by Becker and Chambers (1977) at Bell Laboratories.

11. General

The probability approach to cluster analysis is not often used in practice in view of some computational and assumptional disadvantages. A thorough treatment of this approach is however presented in Mardia, Kent, and Bibby (1979) and Hawkins, Muller, and Ten Krooden (1982).

Hierarchical cluster analysis is based on distances and similarities. Metric distance measures are the most widely used measures and were therefore treated thoroughly in Section 2. Non-metric distance measures are discussed more fully in, for example, Anderberg (1973). As far as similarity coefficients are concerned Everitt (1974) and Anderberg (1973) give a comprehensive review.

Hierarchical cluster analysis is usually performed by agglomerative methods and all the computer programs presented in this book use these methods. On the other hand, so-called divisive methods which stand in contrast to the agglomerative methods are seldom used. The interested reader should consult Lance and Williams (1965) and Edwards and Cavalli–Sforza (1965) in this regard.

Holland and Leinhardt (1981) describe an exponential family of distributions that can be used for analyzing digraph data given in the form of an adjacency matrix. They present a substantive rationale for a general model and they analyze Sampson's data by way of an example.

CHAPTER 6

Multidimensional Scaling

1. Introduction

A data matrix usually contains too much information to absorb at once. The differences between the various rows and columns as well as the interactions between them, are difficult to determine merely by looking at the matrix. However, if the above information can be simplified to representations in one, two or three dimensions at the most, the human eye is usually capable of observing any differences and interactions between rows and columns with the aid of geometrical distance comparisons. This simplifying process is commonly known as multidimensional scaling.

Three types of data matrices will be considered in this chapter. The schematic representation of two-dimensional scaling, which can be made of each of these data matrices according to Greenacre and Underhill (1982), will now be discussed briefly.

A. Profile Data

In this situation we observe the profiles of n cases. The Ability Test data set (Table 1.1) illustrates such a data matrix. A possible two-dimensional scaling is as follows:

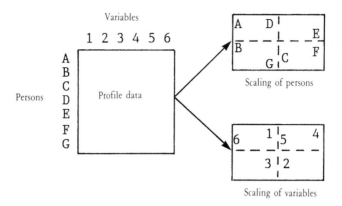

B. Frequency Data

Assume n cases are classified according to two factors, A and B. The data matrix now consists of frequencies or numbers. In this case a possible two-dimensional scaling is:

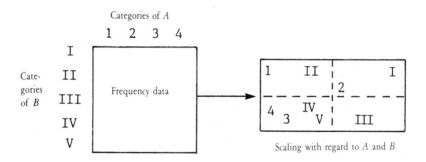

C. Distance or Similarity Data

The information in respect of n cases is often presented by means of a symmetrical distance (dissimilarity) or similarity matrix. Consider the following example. Five political leaders are assessed on a 10-point scale by a number of voters according to the differences between their political views. The combination (leader A, leader B) will therefore be awarded a score of 0 if, according to a particular voter they have the same political views. A score of 9, on the other hand, will indicate a radical difference in political views. Calculating a mean score for every combination of leaders gives rise to a symmetrical distance matrix. In this case a possible two-dimensional scaling is:

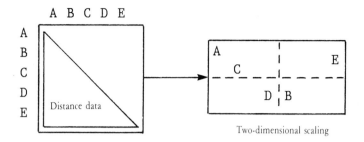

In Section 2 of this chapter we consider the biplot, whereas in Section 3 attention is paid to principal component analysis. These techniques produce multi-dimensional scalings of profile data. Correspondence analysis, which is often based on frequency data (arranged in contingency table form), is considered in Section 4. Classical scaling and principal co-ordinate analysis based on distance and similarity data are dealt with in Section 5, while non-metric scaling is discussed in Section 6. This last-mentioned distance data technique is based only on the ordering of the distances in the distance matrix and does not use the distances as such. INDSCAL, a three-way multi-dimensional scaling technique, is discussed in Section 7. An introduction to Guttman's techniques is given in Section 8 and his facet theory approach is discussed in Section 9. Finally, POSAC a two-dimensional scaling technique for the exploratory analysis of multivariate ordinal data is presented in Section 10.

2. The Biplot

A biplot [see Gabriel (1971) and (1980)] is a graphical representation of the rows and columns of a data set $\mathbf{X}$. The prefix "bi" in "biplot" refers to the joint representation of the rows and columns of the matrix $\mathbf{X}$, and not to the fact that the visual representation is of a two-dimensional nature.

The columns of such a data set $\mathbf{X}$, represent variables and the rows represent cases (e.g. individuals). The data set given in Table 1.1, for example, has 28 rows (the 28 pupils) and 22 columns (the variables). A data set or data matrix $\mathbf{X}$ can be written symbolically as follows:

$$\mathbf{X} = \begin{array}{c} \text{Variables} \quad X_1 \quad X_2 \qquad X_p \quad \textit{Cases} \\ \begin{bmatrix} X_{11} & X_{12} & \cdots & X_{1p} \\ X_{21} & X_{22} & \cdots & X_{2p} \\ \vdots & \vdots & & \vdots \\ X_{n1} & X_{n2} & \cdots & X_{np} \end{bmatrix} \begin{array}{c} 1 \\ 2 \\ \vdots \\ n \end{array} \end{array}$$

In the above notation X_{21} for instance indicates the outcome of variable 1 in respect of the second case.

Assume that $l_1, l_2, \ldots, l_p$ are the eigenvalues (eigenroots) of $\mathbf{X}$ and that $l_1 \geq l_2 \cdots \geq l_p$. Assume further that $\mathbf{U}_1, \mathbf{U}_2, \ldots, \mathbf{U}_n$ and $\mathbf{V}_1, \mathbf{V}_2, \ldots, \mathbf{V}_p$ are respectively the left and right eigenvectors of $\mathbf{X}$. By means of the so-called singular value factorization of a rectangular matrix it now approximately holds that

$$
\mathbf{X} \simeq \mathbf{X}_{(2)} =
\begin{bmatrix}
U_{11} l_1^k & U_{12} l_2^k \\
U_{21} l_1^k & U_{22} l_2^k \\
\vdots & \vdots \\
U_{n1} l_1^k & U_{n2} l_2^k
\end{bmatrix}
\begin{bmatrix}
V_{11} l_1^{1-k} & V_{21} l_1^{1-k} & \cdots & V_{p1} l_1^{1-k} \\
V_{12} l_2^{1-k} & V_{22} l_2^{1-k} & \cdots & V_{p2} l_2^{1-k}
\end{bmatrix}
$$

$$
= \mathbf{AB}', \text{ whereby } 0 \leq k \leq 1.
$$

The above approximation is known as the rank two approximation of $\mathbf{X}$ and usually is satisfactory if $l_3, l_4, \ldots, l_p$ are small in relation to l_1 and l_2. Similarly higher rank approximations can be obtained by including additional columns and rows for the matrices $\mathbf{A}$ and $\mathbf{B}$ respectively.

The corresponding elements of the first and second columns of the matrix $\mathbf{A}$ are the co-ordinates for plotting the n cases. Similarly the corresponding elements of the first and second rows of $\mathbf{B}'$ are the co-ordinates for plotting the p variables. The values l_1^{1-k} and l_2^{1-k} determine the length of the axes.

A measure of how satisfactorily $\mathbf{X}_{(2)}$ approximates the original matrix $\mathbf{X}$, is given by the following goodness of fit criterion:

$$
\frac{l_1 + l_2}{l_1 + l_2 + \cdots + l_p}.
$$

From the expression for $\mathbf{X}_{(2)}$ and the above formula it follows that $\mathbf{X}_{(2)} = \mathbf{X}$ if $l_3 = l_4 = \cdots = l_p = 0$.

A biplot is usually obtained from the matrix $\mathbf{Y}$ with the elements of $\mathbf{Y}$ equalling the deviations of the original observations from the corresponding means:

$$
\mathbf{Y} =
\begin{bmatrix}
X_{11} - \bar{X}_1 & X_{12} - \bar{X}_2 & \cdots & X_{1p} - \bar{X}_p \\
X_{21} - \bar{X}_1 & X_{22} - \bar{X}_2 & \cdots & X_{2p} - \bar{X}_p \\
\vdots & \vdots & & \vdots \\
X_{n1} - \bar{X}_1 & X_{n2} - \bar{X}_2 & \cdots & X_{np} - \bar{X}_p
\end{bmatrix}.
$$

If a biplot is made for rows and columns jointly, $k = 1/2$ is chosen, whereas $k = 0$ is usually chosen if only the columns (variables) are represented. Figure 6.1 illustrates a biplot in respect of four imaginary variables.

If $k = 0$, the lengths OX_1, OX_2, etc. represent the standard deviations of the corresponding variables. The correlation between any two variables X_i and X_j is the cosine of the angle between the lines OX_i and OX_j. From Figure 6.1, for instance, it follows that the standard deviation of X_4 is considerably smaller than that of X_1 or X_3. The correlation between X_1 and X_2 is approximately the cosine of $90°$, i.e. 0. On the same basis it follows that

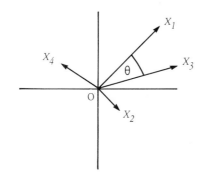

Figure 6.1. Biplot of **Y** in respect of columns (variables).

$$r_{X_1, X_3} = \cos 30° = 0.87 \quad \text{and}$$

$$r_{X_2, X_4} = \cos 190° = -0.99.$$

Therefore, the variables X_1 and X_2 are uncorrelated but there is a high degree of correlation between X_1 and X_3 and between X_2 and X_4.

Suppose we want to obtain a biplot of the 28 observations of the variables X_{13} to X_{20} [the Educational Level III ability test variables as well as the Junior Aptitude Test (JAT) and the Senior Aptitude Test (SAT)] in the data set given in Table 1.1. The following plotting symbols are used:

$A = X_{13}$ (Educational Level III, Number Series)
$B = X_{14}$ (Educational Level III, Figure Analogies)
$C = X_{15}$ (Educational Level III, Pattern Completion)
$D = X_{16}$ (Educational Level III, Classification: Word Pairs)
$E = X_{17}$ (Educational Level III, Verbal Reasoning)
$F = X_{18}$ (Educational Level III, Word Analogies)
$G = X_{19}$ (Junior Aptitude Test: Reasoning)
$H = X_{20}$ (Senior Aptitude Test: Verbal)
1 = Language Group A, Males
2 = Language Group A, Females
3 = Language Group B, Males
4 = Language Group B, Females.

The SAS PROC MATRIX and SAS PROC PLOT programs were used firstly, to calculate co-ordinates for the row and column variables and secondly, to plot these co-ordinates. For the joint representation of the rows and columns $k = 1/2$ was chosen and a matrix **Y** of deviations from the column means calculated. The computer program is as follows:

Input: SAS PROC MATRIX program and SAS PROC PLOT program for obtaining a biplot of a data set

```
PROC    MATRIX ;
        X = 19 19 19 20 17 17 39 21/
            21 15 17 18 18 19 42 22/
            15 18 13 15 18 13 38 18/
                      .
                      .
                      .
            13 18 21 12 13 15 33 14/
            13 20 20 21 15 20 38 21/
            11 13 14 14 15 14 39 12 ;
* X IS A DATA MATRIX. THE NUMBER OF ROWS OF X ARE THE NUMBER OF
  OBSERVATIONS AND THE NUMBER OF COLUMNS OF X THE VARIABLES. ;
        NOBS = NROW(X) ;
        NVAR = NCOL(X) ;
        NVM1 = NVAR-1 ;
* CALCULATE THE MEAN VECTOR FOR THE VARIABLES. THEN CALCULATE THE
  MATRIX (DIFF) WHICH CONTAINS THE DEVIATIONS OF THE OBSERVATIONS
  FROM THE COLUMN MEANS ;
        MEAN = X(+,) #/ NOBS ;
        V_ONE = J(NOBS,1,1) ;
        DIFF = (X - V_ONE*MEAN) ;
* DIFF CAN BE WRITTEN AS U * D_Q * V': COMPUTE U,Q AND V ;
        SVD U Q V DIFF ;
* ARRANGE THE ELEMENTS OF Q IE. THE EIGENVALUES IN ASCENDING ORDER
  AND ARRANGE THE EIGENVECTORS (THE ELEMENTS OF U AND V)
  IN THE SAME ORDER ;
        R = RANK(Q) ;
        TEMP=U ;U(,R)=TEMP ;
        TEMP=Q ;Q(R,)=TEMP ;
        TEMP=V ;V(,R)=TEMP ;
* SET THE ELEMENTS OF E, F AND G TO ZERO ;
        E = J(NOBS,2,0) ;
        F = J(NVAR,2,0) ;
        G = J(2,2,0) ;
* SELECT THE LAST AND SECOND LAST LEFT EIGENVECTOR ;
        E(,1) = U(1:NOBS,NVAR) ;
        E(,2) = U(1:NOBS,NVM1) ;
* SELECT THE LAST AND SECOND LAST RIGHT EIGENVECTOR ;
        F(,1) = V(1:NVAR,NVAR) ;
        F(,2) = V(1:NVAR,NVM1) ;
* SELECT THE LARGEST AND SECOND LARGEST IE. THE LAST AND SECOND
  LAST EIGENVALUE ;
        R = Q(NVAR,1) ;
        S = Q(NVM1,1) ;
* CALCULATE THE GOODNESS OF FIT STATISTIC ;
        SUM = Q(+,) ;
        G_OF_FIT = (R+S) #/ SUM ;
        G_OF_FIT = G_OF_FIT * 100 ;
* CALCULATE THE COORDINATES AND STORE THEM IN THE MATRIX BIPLOT ;
        K = 0.5 ;
        KC = 1-K ;
        G(1,1) = EXP( K * LOG(R) ) ;
        G(2,2) = EXP( K * LOG(S) ) ;
        ABIPLOT = E * G ;
        G(1,1) = EXP( KC * LOG(R) ) ;
        G(2,2) = EXP( KC * LOG(S) ) ;
        BBIPLOT = F * G ;
        BIPLOT = ABIPLOT // BBIPLOT ;
PRINT   BIPLOT G_OF_FIT K ;
OUTPUT  BIPLOT OUT=COORDIN
        (RENAME=(COL1=X COL2=Y)) ;
* DEFINE THE SYMBOLS TO BE USED IN THE PLOTS ;
DATA    COORDIN ;
        SET COORDIN ;
        LENGTH SYMBOL $1 ;
        SYMBOL =
        SUBSTR('44112112213434333321211222ABCDEFGH',_N_,1) ;
* PLOT THE POINTS ;
PROC    PLOT DATA=COORDIN ;
        PLOT Y*X =SYMBOL / VREF=0 HREF=0 VPOS=30 HPOS=50 ;
TITLE   'Biplot of the observations and the variables ;
```

Output:

(i) *Co-ordinates for the rows (first 28 pairs) and columns (last 8 pairs)*

SAS

BIPLOT	COL1	COL2
ROW1	-1.40904	0.469764
ROW2	-1.71669	-0.433265
ROW3	-0.353201	-0.558127
ROW4	1.11023	-1.14073
ROW5	-2.35685	0.327798
ROW6	-0.552231	0.50249
ROW7	-1.40117	-0.878483
ROW8	0.728408	0.925536
ROW9	0.0825417	-1.34952
ROW10	-0.183152	0.937944
ROW11	2.16133	0.875374
ROW12	-1.75163	-0.325469
ROW13	1.96813	-0.354769
ROW14	1.2045	1.54636
ROW15	0.345588	-0.00187498
ROW16	2.13649	-0.793963
ROW17	0.53399	0.905918
ROW18	0.794886	-0.438632
ROW19	1.50674	-0.643398
ROW20	0.565125	-0.271546
ROW21	-0.693138	0.24187
ROW22	-0.309383	1.18742
ROW23	0.146675	0.156051
ROW24	-0.645317	-1.47748
ROW25	-1.43303	-0.0222791
ROW26	0.420036	1.02478
ROW27	-1.17548	0.907051
ROW28	0.275643	-1.31883
ROW29	-1.7808	0.0905062
ROW30	-1.08785	2.05922
ROW31	-1.79563	2.59695
ROW32	-1.75975	-0.135981
ROW33	-2.20055	0.542537
ROW34	-1.7265	1.01119
ROW35	-4.22867	-2.60991
ROW36	-2.06387	0.616135

G_OF_FIT	COL1
ROW1	44.6433

K	COL1
ROW1	0.5

From the computer output (G_OF_FIT = 44.64%) it appears that the measure of how well $X_{(2)}$ fits the data set is fairly low. The biplot is subsequently given.

(ii) *Biplot of the data set in Table* 1.1 *(Variables* X_{13} *to* X_{20}*)*

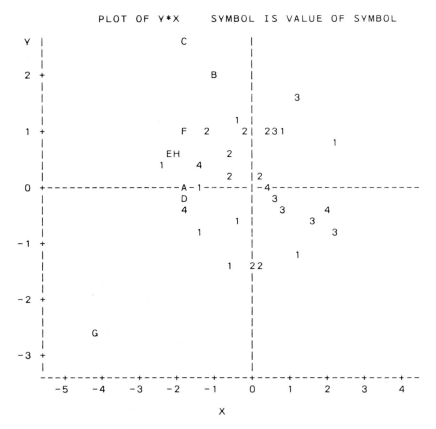

PLOT OF Y*X   SYMBOL IS VALUE OF SYMBOL

NOTE: 1 OBS HIDDEN

The biplot shows considerable variation in the pupils' achievement. Row points (1, 2, 3 and 4) occurring close to the column points $(A, B, C, \ldots, H)$ on the plot, indicate pupils who obtained relatively high scores in the various tests. However no clear distinction is evident between laguage-sex groups.

The fact that points A, D, E, H and F lie relatively closely to one another indicates a fairly high correlation between the above variables and also that the standard deviations of these variables are of the same order. A plot of the column points only, with $k = 0$, was subsequently made with a view to carefully examining the above-mentioned aspects.

The final part of the biplot program given earlier was amended as follows:

Input: SAS PROC PLOT program for plotting the column points only

```
OUTPUT BBIPLOT OUT=COLUMNS
       (RENAME=(COL1=W COL2=Z)) ;
* DEFINE THE SYMBOLS TO BE USED IN THE PLOTS ;
DATA   COLUMNS ;
       SET COLUMNS ;
       LENGTH SYMBOL $1 ;
       SYMBOL=SUBSTR('ABCDEFGH',_N_,1) ;
* PLOT THE VARIABLES ;
PROC   PLOT DATA=COLUMNS ;
       PLOT Z*W=SYMBOL / HAXIS=-27 TO 0 BY 1.5  VREF=0 HREF=0
                         VAXIS=-13.5 TO 13.5 BY 1.5  VPOS=30 HPOS=50
                         VSPACE=30 HSPACE=50 ;
TITLE  'Biplot of the variables (columns)' ;
```

Output:

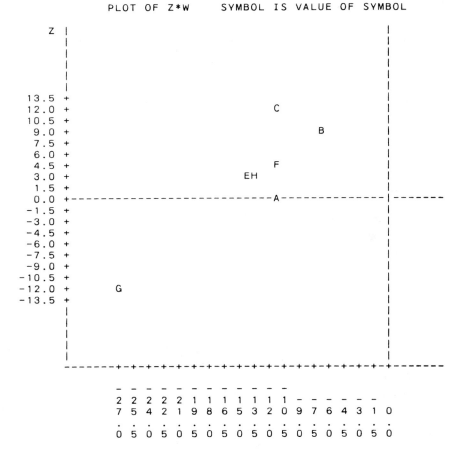

NOTE: 1 OBS HIDDEN

The correlation between B and G, i.e. between X_{14} and X_{19}, is approximately $\cos 70° = 0.342$. From the plot it follows that the standard deviations of variables B and G are, respectively, the smallest and the largest and that the variables A, D, E, H and F have relatively high inter-correlations with approximately equal standard deviations.

3. Principal Component Analysis

Consider again the data matrix $\mathbf{X}$ of the previous section where p measurements $X_1, X_2, \ldots, X_p$ were made on each of n randomly selected cases. Principal component analysis involves a linear transformation of these variables to a set of new variables $Y_1, Y_2, \ldots, Y_p$ (called principal components or factors) which have the following useful characteristics:

(a) The p principal components, each of which is a linear combination of the original variables, are uncorrelated.
(b) The first principal component

$$Y_1 = a_{11} X_1 + a_{21} X_2 + \cdots + a_{p1} X_p$$

explains the largest percentage of the variation in the original multivariate data set, whereas the second principal component

$$Y_2 = a_{12} X_1 + a_{22} X_2 + \cdots + a_{p2} X_p$$

explains the second largest percentage of the variation, etc.

Usually the first few factors jointly explain the variation present in the original p-dimensional data set to a great extent, whereas the remaining factors make only a slight contribution. Suppose, for instance, that the first factor explains 80% or more of the variation in an eight-variate data set. The eight-dimensional data set can then be represented one-dimensionally by the Y_1 values without too much loss of information. In this case the coefficients $a_{11}, a_{21}, \ldots, a_{81}$ in the linear combination Y_1 indicate the relative importance of each of the original eight variables in relation to factor Y_1.

If the principal components are based on the sample covariance matrix $\mathbf{S}$, the coefficients a_{ij} in $Y_1, Y_2, \ldots, Y_p$ are known as unstandardized coefficients, whereas if the principal components are based on the sample correlation matrix $\mathbf{R}$ the coefficients are known as standardized coefficients. Determining principal components amounts to determining the eigenvalues and eigenvectors of $\mathbf{S}$ (or $\mathbf{R}$). These eigenvalues and eigenvectors are usually obtained by means of a singular value factorization of a symmetric matrix. The elements of the first eigenvector $(a_{11}, a_{21}, \ldots, a_{p1})$ are the coefficients in Y_1, whereas the value l_1 of the corresponding eigenvalue is the variance of the first principal component. Generally the elements of the ith eigenvector $(a_{1i}, a_{2i}, \ldots, a_{pi})$ are the coefficients in Y_i with corresponding root l_i. The percentage variation explained by principal component Y_i is

$$\frac{l_i}{l_1 + l_2 + \cdots + l_p} \times 100\%.$$

A. Two-Dimensional Scaling of Variables

Denote the correlation between variable X_i and component Y_j by the symbol r_{X_i, Y_j}. It can then be shown that

$$r_{X_i, Y_j} = a_{ij}\sqrt{l_j}/s_i; \qquad i, j = 1, 2, \ldots, p$$

if the analysis is based on $\mathbf{S}$, where s_i and $\sqrt{l_j}$ are the sample standard deviations of X_i and Y_j respectively. If it is assumed that Y_1 and Y_2 explain a reasonable amount of the variation in the original data set, a graphical representation of the p points $(r_{X_i, Y_1}; r_{X_i, Y_2})$ can assist the researcher in placing the original variables $X_1, X_2, \ldots, X_p$ in fairly homogeneous groups. If the researcher wants to limit his investigation to a smaller number of variables, he now chooses a subset from each of the homogeneous groups of variables. This application of principal component analysis can be regarded as an exploratory technique through which better insight is obtained into the underlying correlation structure of a particular data set.

In order to illustrate this procedure the SAS PROC FACTOR program was used to carry out a principal component analysis on the 18 ability test variables with regard to the males from Language Group B ($n = 663$).

Input: SAS PROC FACTOR program for principal component analysis

```
DATA    ABILITY ;
INPUT   X1-X18 ;
CARDS ;
21 22 22 15 23 23 16 12 16 15 13 14 19 17 16 18 19 18
13 18 21 16 17 15 11 12 11 09 11 11 16 13 14 13 15 18
18 12 09 09 15 17 09 05 03 12 07 07 12 10 10 13 10 12
           .
           . REST OF DATA OF 663 MALES FROM LANGUAGE GROUP B
           .

;
PROC    FACTOR METHOD=PRIN NFACT=2 CORR EIGENVECTORS PLOT ;
VAR     X1-X18 ;
TITLE   'Ability test scores - Language Group B Males' ;
```

Output: Some relevant output of a principal component analysis

(i) *Eigenvalues* (l_i)

	1	2	3	4	5	6
EIGENVALUE	7.424953	1.448882	1.284564	1.077626	0.831112	0.698523
DIFFERENCE	5.976070	0.164319	0.206938	0.246514	0.132589	0.038922
PROPORTION	0.4125	0.0805	0.0714	0.0599	0.0462	0.0388
CUMULATIVE	0.4125	0.4930	0.5644	0.6242	0.6704	0.7092

	7	8	9	10	11	12
EIGENVALUE	0.659601	0.548830	0.526795	0.514188	0.481508	0.423663
DIFFERENCE	0.110770	0.022035	0.012608	0.032680	0.057844	0.003381
PROPORTION	0.0366	0.0305	0.0293	0.0286	0.0268	0.0235
CUMULATIVE	0.7458	0.7763	0.8056	0.8342	0.8609	0.8845

	13	14	15	16	17	18
EIGENVALUE	0.420282	0.383168	0.361476	0.343583	0.315995	0.255252
DIFFERENCE	0.037114	0.021692	0.017893	0.027588	0.060743	
PROPORTION	0.0233	0.0213	0.0201	0.0191	0.0176	0.0142
CUMULATIVE	0.9078	0.9291	0.9492	0.9683	0.9858	1.0000

(ii) *Coefficients a_{ij} of the first two components*

	1	2
X1	0.23292	0.16766
X2	0.23921	0.00237
X3	0.22907	0.19965
X4	0.23401	-0.28087
X5	0.27783	-0.04121
X6	0.20660	-0.23106
X7	0.22413	0.31905
X8	0.22068	0.06293
X9	0.21241	0.29577
X10	0.23373	-0.30350
X11	0.28529	0.01779
X12	0.23824	-0.30383
X13	0.22498	0.31915
X14	0.21679	0.04168
X15	0.20564	0.34243
X16	0.23312	-0.25782
X17	0.28379	0.06501
X18	0.22320	-0.35851

(iii) *Correlations between variables and the first two components (factor pattern matrix)*

	$a_{i1}\sqrt{l_1}$	$a_{i2}\sqrt{l_2}$
	FACTOR1	FACTOR2
X1	0.63468	0.20181
X2	0.65182	0.00285
X3	0.62418	0.24031
X4	0.63765	-0.33808
X5	0.75704	-0.04960
X6	0.56296	-0.27813
X7	0.61072	0.38404
X8	0.60132	0.07575
X9	0.57879	0.35602
X10	0.63689	-0.36532
X11	0.77738	0.02141
X12	0.64916	-0.36572
X13	0.61303	0.38416
X14	0.59073	0.05018
X15	0.56035	0.41219
X16	0.63522	-0.31033
X17	0.77328	0.07825
X18	0.60818	-0.43153

(iv) *Graphical representation of* r_{X_i, Y_1} *against* r_{X_i, Y_2}

```
        PLOT OF FACTOR PATTERN FOR FACTOR1    AND FACTOR2

                              FACTOR1
                                 1
                                .9      II

                               .K
          I                   E    Q
                                .7
                    DP          B        A
                  R            .6NH     C   IG      III
                     F          .5                 O
                                .4
                                .3
                                .2
                                                                F
                                .1                              A
 -1 -.9-.8-.7-.6-.5-.4-.3-.2-.1  0 .1 .2 .3 .4 .5 .6 .7 .8 .9 1.0T
                               -.1                              O
                                                                R
                               -.2                              2
                               -.3
                               -.4
                               -.5
                               -.6
                               -.7
                               -.8
                               -.9
                                -1


      X1=A     X2=B    X3=C    X4=D     X5=E   X6=F    X7=G    X8=H   X9=I

     X10=D   X11=K   X12=D   X13 =G   X14=N  X15=O   X16=P   X17=Q  X18=R
```

Part (i) of the output shows the values of the eigenvalues as well as the percentage variation explained by each component. Note that the first two components explain 49% of the variaton in the 18-dimensional data set. Parts (ii) and (iii) of the output respectively are the coefficients a_{ij} of the eigenvectors and the factor pattern matrix. These last numbers are obtained by multiplying the elements of the ith eigenvector with the square root of the ith eigenvalue, i.e. $a_{ij}\sqrt{l_i}$ (cf. the biplot, Section 2). Since the eigenvalues and eigenvectors have been determined on the basis of the correlation matrix it follows that $r_{X_i, Y_j} = a_{ij}\sqrt{l_i}$. Part (iv) of the computer output is a graphical representation of the original 18 variables with regard to their correlations with components 1 and 2. This representation is obtained by plotting the r_{X_i, Y_1} against r_{X_i, Y_2}, $i = 1, \ldots, 18$.

From this graphical representation it follows that the test for word analogies and classification (word pairs) for Educational Levels I, II and III $(X_4, X_6, X_{10}, X_{12}, X_{16}, X_{18})$ are a fairly homogeneous group of variables. In Output (iv) they are identified as Group I. Group II represents the tests for figure analogies and verbal reasoning for the three educational levels $(X_2, X_5, X_8, X_{11}, X_{14}$ and $X_{17})$ whereas Group III contains all the tests in number series and pattern completion $(X_1, X_3, X_7, X_9, X_{13}$ and $X_{15})$.

The points (variables) in Output (iv) which are situated closely to the unit circle, are well explained by the first two components. Points situated closer to the origin are explained to a lesser extent by these two principal components. It is interesting to note that the variables in Group II are to a great extent explained by component 1 since they are situated on the first component axis and are relatively far from the origin. In Output (iv) Group I refers to the cluster (R, D, P, F), Group II to the cluster (K, E, Q, B, N, H) and Group III to the cluster (A, C, I, G, O).

B. Two-Dimensional Scaling of Cases

It is possible to calculate p principal component scores for each p-variate observation vector, namely

$$Y_1 = a_{11}X_1 + a_{21}X_2 + \cdots + a_{p1}X_p$$
$$Y_2 = a_{12}X_1 + a_{22}X_2 + \cdots + a_{p2}X_p$$
$$\vdots$$
$$Y_p = a_{1p}X_1 + a_{2p}X_2 + \cdots + a_{pp}X_p.$$

If the p measurements $(X_1, X_2, \ldots, X_p)$ of a specific case are substituted in the first two components Y_1 and Y_2, the score (Y_1, Y_2) is obtained. If Y_1 is plotted against Y_2 a two-dimensional scaling of the n cases is obtained.

SAS PROC PRINCOMP may be used to represent this two-dimensional scaling graphically. We illustrate this procedure by using the data for the ability tests X_1 to X_{18} (see Table 1.1).

Input: SAS PROC PRINCOMP program for two-dimensional scaling

```
DATA    ABILITY ;
INPUT   PUPIL: $1. X1-X18 ;
CARDS ;
 1   19 21 21 18 20 21 15 14 15 13 15 16 19 19 19 20 17 17
 2   21 20 15 24 22 18 11 18 16 19 14 17 21 15 17 18 18 19
 3   18 19 16 18 18 23 11 13 13 15 11 11 15 18 13 15 18 13
         .
         .
         .
 Q   22 17 21 17 17 22 10 14 16 16 13 08 13 18 21 12 13 15
 R   21 18 20 23 21 22 08 15 09 17 11 13 13 20 20 21 15 20
 S   21 22 19 20 18 17 11 15 12 14 11 10 11 13 14 14 15 14
 ;
PROC    PRINCOMP OUT=PRIN N=2 ;
        VAR X1-X18 ;
PROC    PLOT DATA=PRIN ;
        PLOT PRIN2*PRIN1=PUPIL / VPOS=15 HPOS=55 ;
PROC    PRINT ;
        ID PUPIL ;
        VAR PRIN1 PRIN2 ;
```

Output: Two-dimensional scaling of pupils 1 to 28 from Table 1.1

(i) *Eigenvalues and goodness of fit*

	EIGENVALUE	DIFFERENCE	PROPORTION	CUMULATIVE
PRIN1	8.507366	6.798379	0.472631	0.472631
PRIN2	1.708988		0.094944	0.567575

(ii) *Eigenvectors*

	PRIN1	PRIN2
X1	0.230360	-.165457
X2	0.237289	-.218377
X3	0.235192	-.040606
X4	0.244164	-.233966
X5	0.259075	0.220926
X6	0.258226	-.217103
X7	0.104438	0.565515
X8	0.267891	-.073176
X9	0.239365	-.055890
X10	0.241385	-.369548
X11	0.264733	0.187372
X12	0.250140	0.143526
X13	0.168612	0.472451
X14	0.158305	-.016712
X15	0.247166	0.065446
X16	0.235744	-.033691
X17	0.287031	0.150983
X18	0.241793	0.018478

(iii) 28 *Principal component scores*

PUPIL	PRIN1	PRIN2
1	2.2457	1.7257
2	2.8747	0.5133
3	-0.3995	-0.1413
4	-3.6982	-1.0410
5	5.3014	1.7701
6	1.9629	0.6322
7	2.4523	0.4914
8	-0.2739	-2.0386
9	-1.3303	-0.9491
A	2.1149	-1.2677
B	-4.2446	0.1669
C	2.2975	1.6420
D	-2.7166	-1.7295
E	-2.0710	1.6061
F	-1.3814	1.8696
G	-6.4938	0.3923
H	-1.4971	0.5164
I	-2.1537	1.3001
J	-4.8190	0.3519
K	-2.3972	-0.0376
L	3.0560	-1.4892
M	2.5615	-0.6723
N	2.7286	-2.4234
O	0.8070	2.3004
P	3.9123	-0.3739
Q	0.0815	-0.9249
R	1.8594	-1.3237
S	-0.7793	-0.8661

(iv) *Two-dimensional scaling of* 28 *pupils*

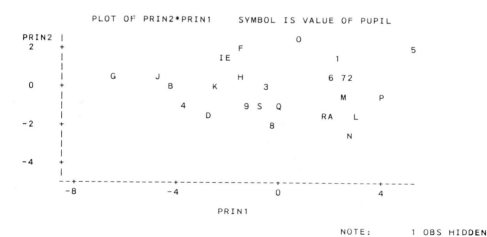

From the two-dimensional display it appears that points *G* (pupil number 16) and 5 (pupil number 5) are the most extreme. This is confirmed by their overall scores as given in Table 1.1 which reveal that they obtained the lowest and highest total scores, respectively.

4. Correspondence Analysis

Correspondence analysis is a technique for simultaneously representing the rows and columns of a two-way classification table. In order to apply this technique the categories of the row and column variables do not have to be mutually exclusive, which makes it particularly appropriate for analyzing two-way tables where a respondent is able to indicate more than one choice or category. The only further requirement is that the various cell entries in the table should be non-negative.

Consider the following $I \times J$ table where $J \leqq I$:

| | Characteristic B (columns) | | | | |
	1	2	...	J	Row total
Characteristic A (rows) 1	f_{11}	f_{12}	...	f_{1J}	r_1
2	f_{21}	f_{22}	...	f_{2J}	r_2
:	:	:		:	:
I	f_{I1}	f_{I2}	...	f_{IJ}	r_I
Column total	c_1	c_2	...	c_J	n

The graphical representation of the rows and columns of this table is based on a singular value factorization (see Section 2) of a matrix $\mathbf{G}$ where

$$
\mathbf{G} = \begin{bmatrix} \dfrac{f_{11}}{\sqrt{r_1 c_1}} & \dfrac{f_{12}}{\sqrt{r_1 c_2}} & \cdots & \dfrac{f_{1J}}{\sqrt{r_1 c_J}} \\[2ex] \dfrac{f_{21}}{\sqrt{r_2 c_1}} & \dfrac{f_{22}}{\sqrt{r_2 c_2}} & \cdots & \dfrac{f_{2J}}{\sqrt{r_2 c_J}} \\[2ex] \vdots & \vdots & & \vdots \\[2ex] \dfrac{f_{I1}}{\sqrt{r_I c_1}} & \dfrac{f_{I2}}{\sqrt{r_I c_2}} & \cdots & \dfrac{f_{IJ}}{\sqrt{r_I c_J}} \end{bmatrix}.
$$

Analogous to the biplot in Section 2

$$
\mathbf{G} \cong \mathbf{G}(2) = \begin{bmatrix} U_{12} l_2^k & U_{13} l_3^k \\ U_{22} l_2^k & U_{23} l_3^k \\ \vdots & \vdots \\ U_{I2} l_2^k & U_{I3} l_3^k \end{bmatrix} \begin{bmatrix} V_{12} l_2^{1-k} & V_{22} l_2^{1-k} & \cdots & V_{J2} l_2^{1-k} \\ V_{13} l_3^{1-k} & V_{23} l_3^{1-k} & \cdots & V_{J3} l_3^{1-k} \end{bmatrix}
$$

$$
= \mathbf{AB'}.
$$

Note that the first eigenvectors U_1 and V_1 and the corresponding eigenvalue l_1 do not appear in AB' since it corresponds to the so-called trivial solution. The rank two approximation is evaluated by means of the following goodness of fit criterion:

$$\frac{l_2 + l_3}{l_2 + l_3 + \cdots + l_J}.$$

In order to illustrate this technique, we consider the following contingency table in which the 2682 pupils involved in the project concerning the ability tests have been classified according to language-sex group and proposed university study field. In Table 6.1 the following abbreviations apply:

A-male   = Language Group A, males
A-female = Language Group A, females
B-male   = Language Group B, males
B-female = Language Group B, females.

We will now discuss the use of the program GRAPHICS (a modification of a program developed by N. Tabet of the University of Paris [see du Toit et al. (1982)] with a view to illustrating correspondence analysis. The program for a biplot as given in Section 2 can also be used with a few slight amendments to represent the rows and columns of a contingency table graphically in two dimensions.

The computer input in the file GRAPHICS DATA is as follows:

Input: GRAPHICS DATA input for correspondence analysis

```
                                                                    Section
'Ability test : Language-Sex against University Study Field'           1
4   9   0   2                                                          2
1  1  1  1                                                             3
'MALA'  'FELA'  'MALB'  'FELB'                                         4
12114                                                                 5
   *                                                                  6
'MEDI'    33          5          37          10                       7
'ENG'     61          1          29           0                       7
'NATS'    35         38          29          35                       7
'ARTS'    68         70          41         102                       7
'LAW'     26          1           5           1                       7
'COMM'    59         13          76           8                       7
'MDIP'     5         24           3          59                       7
'EDIP'    21         64           6          65                       7
'NO'     581        198         437         436                       7
```

The data matrix is normally entered in such a way that the number of columns represents the smallest dimensions (in this case, four).

In the program GRAPHICS, so-called supplementary rows (or columns) can also be specified. These are the rows and columns of the data matrix which are not included in the rank two (or higher) approximation of G but which can nevertheless be represented graphically.

The order of the input consists of seven subsections, namely

Table 6.1. Contingency table of 2682 pupils classified according to language-sex group and university study field

University study field	Language-sex				Total
	A-male	A-female	B-male	B-female	
Medicine	33	5	37	10	85
Engineering	61	1	29	0	91
Natural sciences	35	38	29	35	137
Arts	68	70	41	102	281
Law	26	1	5	1	33
Commerce and administration	59	13	76	8	156
Medical diploma	5	24	3	59	91
Education diploma	21	64	6	65	156
No further university study	581	198	437	436	1652
	889	414	663	716	2682

Section 1—Title
Section 2—Parameters
Section 3—Status of variables (columns)
Section 4—Names of variables (columns)
Section 5—Parameters for graphical output
Section 6—Data format
Section 7—Data matrix.

Particulars of each subsection are the following.

A. Section 1

Alphanumerical title of the analysis. The title should consist of 80 characters at the most (including the spaces between words).

B. Section 2

The parameters JCOL, IROWS, SROWS, NAXES. The values of these parameters can be entered in free format. Leave at least one space between numbers.

JCOL   = Total number of columns of data matrix
IROWS = Total number of rows of data matrix
SROWS = Number of supplementary rows.

The supplementary rows always form the final number of rows of the data matrix.

NAXES = Number of axes ("factors") (must be ≤ 7).

C. Section 3

In this section the status of each variable (column) is described by the number 0, 1 or 2, the order of which should correspond to that according to which the columns are entered.
    If the status value

= 0, the particular column is excluded from the analysis
= 1, the particular column is included in the analysis
= 2, the particular column is regarded as a supplementary column.

D. Section 4

The column names. Each column name may consist of four characters at the most and is placed between '' marks. Leave at least one space between names.

E. Section 5

It is possible to obtain as many as eight different plots by means of one run of the computer program GRAPHICS. For each graphical representation the numbers *abcde* are keyed in. For the first graphical representation start with column 1, for the next graphical representation with column 7, then column 13, column 19, etc. where:

a = number of vertical axis
b = number of horizontal axis
c = parameter for the rows (individuals)
   = 0, no graphical representation of the rows
   = 1, representation of rows included in analysis
   = 2, representation of supplementary rows
   = 3, representation of all the rows
d = parameter for the columns (variables)
   = 0, 1, 2 or 3 as for the rows
e = the number of characters of the row names and column names that should appear on the graphical representation ($1 \leq e \leq 4$).

F. Section 6

Input format for the data matrix. The symbol * means free format. Leave at least one space between the elements of the data matrix if free format is used.

G. Section 7

The data matrix (IROWS × JCOL) is keyed in row by row. The data in each row are preceded by a row name (four characters at the most). For data in free format the row names are placed between ' ' marks.

The computer output of the program GRAPHICS appears in the output file GRAPHICS OUTPUT. Part of the output is as follows:

Output: GRAPHICS OUTPUT

(i) *Bar chart of the eigenvalues and goodness of fit*

```
  THE EIGENVALUES              VAL(1)=0.99999893
  ------------------------------------------------------------------------------------------------------
  /NUM /ITER / EIGENVALUE / PERCENT /  CUMUL  /*/   HISTOGRAM OF THE EIGENVALUES (MOMENTS OF INERTIA)
  ------------------------------------------------------------------------------------------------------
  /  2 /   1 / 0.15375531 /  83.440 /  83.440 /*/*..................../....................../...............
  /  3 /   1 / 0.01578250 /   8.565 /  92.004 /*/*******
  /  4 /   4 / 0.01473371 /   7.996 / 100.000 /*/*******
```

From the computer output it follows that the goodness of fit equals 92.0%. This means that the rank two approximation of the two-way table is quite satisfactory.

The co-ordinates for plotting the rows are given by $(a_{11}, a_{12}), (a_{21}, a_{22}), \ldots,$ (a_{I1}, a_{I2}) where a_{ij} indicates a typical element of $\mathbf{A}$. Similarly it follows that the co-ordinates for plotting the columns are given by $(b_{11}, b_{12}), (b_{21}, b_{22}), \ldots,$ (b_{J1}, b_{J2}). These co-ordinates are given in the next section of the computer output.

(ii) *Coordinates for the plotting of the columns and rows*

```
----*----*--------------------*--------------------*-----------------
 J1 /NAME/ QLT MASS INR/   1#F   COR CTR/   2#F   COR CTR/
----*----*--------------------*--------------------*---------------
   1/MALA/ 987   331 212/   309   814 206/  -142   173 427/
   2/FELA/ 854   154 286/  -536   844 289/    58    10  33/
   3/MALB/ 996   247 223/   363   793 212/   184   203 529/
   4/FELB/ 878   267 279/  -409   874 292/   -26     4  12/
----*----*--------------------*--------------------*-------------
   / K= 2682.       1000/          1000/              1000/
----*----*--------------------*--------------------*-------------
```

```
----*----*--------------------*--------------------*-----------------
 I1 /NAME/ QLT MASS INR/   1#F   COR CTR/   2#F   COR CTR/
----*----*--------------------*--------------------*---------------
   1/MEDI/1000    32  51/   506   867  53/   198   133  79/
   2/ENG /  962    34 141/   809   853 144/  -289   109 180/
   3/NATS/  585    51  34/  -248   513  21/    93    72  28/
   4/ARTS/  986   105  90/  -394   983 106/   -22     3   4/
   5/LAW /  940    12  65/   689   487  38/  -664   453 345/
   6/COMM/  975    58 141/   582   759 128/   311   216 356/
   7/MDIP/  898    34 191/  -964   897 206/   -29     1   2/
   8/EDIP/  913    58 253/  -855   913 277/     4     0   0/
   9/NO  /  682   616  34/    82   666  27/   -12    16   6/
----*----*--------------------*--------------------*-------------
   / K= 2682.       1000/          1000/              1000/
----*----*--------------------*--------------------*-------------
```

The QLT column in the computer output gives an indication of the "quality" of the representation of a row (or column). This number is obtained through summation of the "correlations" between the particular element and the respective factors (axes). For the row Medicine the COR scores on the first two factors already give the maximum total of 1000, whereas for Natural Sciences the COR scores on the first two factors render a QLT score of only $(513 + 72) = 585$. This indicates that in a plot of the first two factors only, the quality of the representation of the row point Medicine is very high, whereas the quality of the representation of the row point Natural Sciences is not as high (but nevertheless acceptable). If a row point (individual) therefore has higher COR scores on factors other than the first two, a graphical representation of the first two factors would produce a row point indicating a low-quality score in respect of this representation.

A similar analysis can also be carried out in respect of the column variables.

In the graphical representation below the first factor $(1 \# F)$ is associated with the horizontal axis and the second factor $(2 \# F)$ with the vertical axis.

From this plot it appears that girls of both language groups to a greater extent give preference to diploma training than do boys. The analysis also indicates that boys to a greater extent prefer the so-called professional study fields (e.g., medicine, engineering, law). It is interesting to note that girls of both language groups show a marked preference for the basic sciences compared with the boys. With regard to language differences it would appear that boys from Language Group B show a greater preference for medicine and commerce and administration than boys from Language Group A, whereas the reverse appears to apply in the case of engineering and law.

(iii) *Plot of row and column points*

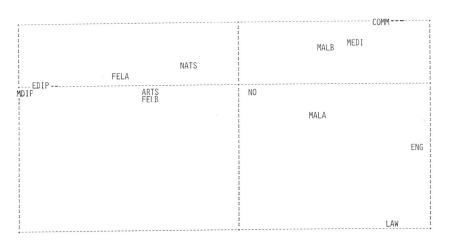

5. Classical (Metric) Scaling

Assume that a distance or dissimilarity matrix $\mathbf{D}$ is observed and that some unknown configuration of points in an Euclidean space of unknown dimension gave rise to this matrix $\mathbf{D}$. Classical or metric scaling attempts to obtain a configuration $\mathbf{X}^*$ in as low a dimension as possible, with distances between the n vectors in the configuration as close to the observed distances as possible. Classical scaling is therefore, in some sense, the inverse of the process of calculating $\mathbf{D}$ from $\mathbf{X}$.

The algorithm used for classical scaling is based on results of Young and Householder (1938), Torgerson (1958) and Gower (1966) and involves the following steps:

(i) Calculate the matrix $\mathbf{A} = \{a_{ij}\}$ where

$$a_{ij} = -\tfrac{1}{2}d_{ij}^2; \qquad i, j = 1, 2, \ldots, n.$$

(ii) Calculate the double-centered matrix $\mathbf{B} = \{b_{ij}\}$ with

$$b_{ij} = a_{ij} - \bar{a}_{i\cdot} - \bar{a}_{\cdot j} + \bar{a}_{\cdot\cdot}; \qquad i, j = 1, 2, \ldots, n,$$

where $\bar{a}_{i\cdot}$, $\bar{a}_{\cdot j}$ and $\bar{a}_{\cdot\cdot}$, respectively, represent the mean of the ith row, the mean of the jth column and the mean of all the elements of $\mathbf{A}$.

(iii) If $\mathbf{D}$ is an Euclidean distance matrix and $\mathbf{B}$ is positive semi-definite it can be proved that

$$\mathbf{B} = \mathbf{XX'}.$$

In practice therefore, if $l_1, l_2, \ldots, l_p$ are the positive eigenvalues of $\mathbf{B}$ with corresponding eigenvectors $\mathbf{V}_1, \mathbf{V}_2, \ldots, \mathbf{V}_p$ the $n \times k$ matrix

$$\mathbf{X}^* = (\sqrt{l_1}\,\mathbf{V}_1, \sqrt{l_2}\,\mathbf{V}_2, \ldots, \sqrt{l_k}\,\mathbf{V}_k), \qquad k \leq p,$$

forms a k-dimensional configuration of the n cases.

If l_1 is much larger than $l_2, l_3, \ldots, l_p$ a one-dimensional representation produces a good scaling of $\mathbf{D}$. In the case where l_1 and l_2 are much larger than the other eigenvalues, a two-dimensional scaling is meaningful, etc. It is important however, always to choose k smaller than the number of positive eigenvalues.

The scaling that is obtained by applying steps (ii) to (iii) on a similarity matrix $\mathbf{C}$, is known as principal co-ordinate analysis. In this regard see Mardia et al. (1979).

It is evident classical scaling can also be applied to an observed data matrix $\mathbf{X}$ where it is of importance to represent this matrix in one or two dimensions. In view of the above algorithm it is necessary to use either the Euclidean metric or its standardized form in the calculation of $\mathbf{D}$.

To illustrate the foregoing scaling method consider the dissimilarity matrix given in Table 6.2.

Table 6.2. Dissimilarities between 9 occupations

1.	0.00								
2.	3.06	0.00							
3.	2.14	3.95	0.00						
4.	3.21	3.95	3.03	0.00					
5.	3.51	4.17	2.82	3.33	0.00				
6.	4.40	3.77	3.86	4.14	3.60	0.00			
7.	3.64	3.69	3.47	3.90	3.56	2.53	0.00		
8.	3.12	4.13	2.72	2.58	3.68	4.17	4.10	0.00	
9.	3.73	4.05	3.31	3.46	3.29	3.92	3.59	3.72	0.00

The elements of this matrix are the means of the opinions of 62 persons as to how divergent nine different occupations are.

The occupations are:

1. Preacher
2. Surgeon
3. Teacher at primary school
4. Journalist
5. Policeman
6. Plumber
7. Farmer
8. Actor
9. Bank clerk

Each of the 36 pairwise comparisons of the 9 occupations were evaluated by the 62 respondents on a scale ranging from 0 (the two occupations are similar) to 5 (the two occupations are totally dissimilar).

The SAS PROC MATRIX program was used to do the classical scaling and the first two vectors of X^* were plotted by using the PLOT procedure.

Input: SAS PROC MATRIX program for performing a classical scaling on the dissimilarity matrix of 9 occupations

```
PROC    MATRIX ;
D= 0.00 3.06 2.14 3.21 3.51 4.40 3.64 3.12 3.73 /
   3.06 0.00 3.95 3.95 4.17 3.77 3.69 4.13 4.05 /
   2.14 3.95 0.00 3.03 2.82 3.86 3.47 2.72 3.31 /
   3.21 3.95 3.03 0.00 3.33 4.14 3.90 2.58 3.46 /
   3.51 4.17 2.82 3.33 0.00 3.60 3.56 3.68 3.29 /
   4.40 3.77 3.86 4.14 3.60 0.00 2.53 4.17 3.92 /
   3.64 3.69 3.47 3.90 3.56 2.53 0.00 4.10 3.59 /
   3.12 4.13 2.72 2.58 3.68 4.17 4.10 0.00 3.72 /
   3.73 4.05 3.31 3.46 3.29 3.92 3.59 3.72 0.00 ;
            N = NROW(D) ;
            E = J(1,N,1.0) ;
            A = -0.5 * (D#D) ;
            C1 = 1.0 #/ N ;
            C2 = C1 * C1 ;
            GKR = C1 * A(+,) ;
            GG = C2 * SUM(A) ;
            B = A - (E'*GKR) - (GKR'*E) + (GG*(E'*E)) ;
            EIGEN  W V B ;
  * STANDARDIZE V SO THAT V'*V=W ;
            VN = V * HALF(DIAG(W)) ;
  PRINT   D B W VN ;
  OUTPUT VN OUT=A ;
  DATA    A ;
  SET     A ;
  LENGTH S S1 ;
            S = _N_ ;
  PROC    PLOT ;
            PLOT COL2*COL1=S / VREF=0 HREF=0 VPOS=30 HPOS=50 ;
```

Output: Some relevant output

(i) *Eigenvalues of* **B**

W	COL 1
ROW1	15.4544
ROW2	10.6958
ROW3	7.08561
ROW4	6.52556
ROW5	4.95509
ROW6	3.57874
ROW7	1.98255
ROW8	0.821644
ROW9	6.1455E-14

Since $(l_1 + l_2)/(l_1 + l_2 + \cdots + l_9) = 0.512$ a two-dimensional configuration should give some information on the Euclidean structure underlying the dissimilarity matrix **D**.

(ii) *Two-dimensional configuration* $(\sqrt{l_1}\,\mathbf{V}_1, \sqrt{l_2}\,\mathbf{V}_2)$

VN	COL 1	COL 2
ROW1	-1.09931	-1.3406
ROW2	0.773266	-2.39454
ROW3	-0.998378	0.250529
ROW4	-1.26321	0.353279
ROW5	-0.021418	1.27202
ROW6	2.31579	0.486216
ROW7	1.8112	0.140189
ROW8	-1.55258	0.202285
ROW9	0.0346418	1.03062

(iii) *Plot of two-dimensional configuration*

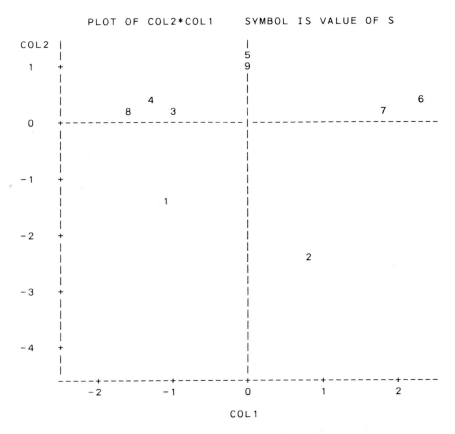

The two-dimensional configuration suggests three homogeneous occupational groups, namely farmer and plumber; policeman and bank clerk; actor, teacher and journalist. The occupations of surgeon and preacher are dissimilar not only from each other but also from the rest. Some care must be exercised in reaching these conclusions since the two dimensional plot is by far not an exact reconstruction of the underlying profile matrix resulting in **D**. The interpretation of the two dimensional scaling of the occupation dissimilarity matrix will be discussed more fully in the next section.

6. Non-Metric Scaling

According to Mardia et al. (1979) the aim of metric scaling is to reconstruct the "true" configuration in p dimensions given an observed distance matrix **D** whose elements are of the form

$$d_{ij} = \delta_{ij} + \varepsilon_{ij}.$$

The ε_{ij} represent errors in measurement and dimensionality and δ_{ij} is the interpoint distance between the reconstructed vectors. Shepard (1962) proposed a model which only requires the observed dissimilarities to be monotonically related to the true Euclidean distances. In the above notation he assumed therefore that

$$d_{ij} = f(\delta_{ij} + \varepsilon_{ij})$$

where f is an unknown monotone increasing function. The only information that can be used to find the "true" configuration is the rank order of the d_{ij}.

Suppose that the elements of the observed matrix $\mathbf{D}$ are ordered in the following way

$$d_1 < d_2 < \cdots < d_m; \qquad m = n(n-1)/2.$$

If $\mathbf{X}^*$ is a chosen $\mathbf{X}$ configuration in k dimensions and $\mathbf{D}^*$ the corresponding distance matrix (usually the Minkovsky metric is used, see Chapter 5) then $d_1^*, d_2^*, \ldots, d_m^*$ correspond to the $d_1, d_2, \ldots, d_m$ values. A third set of distances $\hat{d}_1, \hat{d}_2, \ldots, \hat{d}_m$ known as rank images or disparities, are now computed to be as nearly equal to $d_1^*, d_2^*, \ldots, d_m^*$ as possible but subject to the constraint that they are monotonically related to $d_1, d_2, \ldots, d_m$. In other words $\hat{d}_1 \leq \hat{d}_2 \leq \cdots \leq \hat{d}_m$. The calculation of the disparities is known as performing a monotone regression of the d_i^* on the d_i.

In order to evaluate the fit for a given $\mathbf{X}^*$ configuration Kruskal's (1964) STRESS-formula is usually used

$$S_1(\mathbf{X}^*) = \left\{ \sum_{i=1}^{m} (d_i^* - \hat{d}_i)^2 \Big/ \sum_{i=1}^{m} (d_i^*)^2 \right\}^{1/2}.$$

If $S_1(\mathbf{X}^*) = 0$ it means that a monotone relationship exists between the elements of $\mathbf{D}$ and $\mathbf{D}^*$. In view of all the errors in measurement involved and the fact that k is usually chosen to be small, this will hardly ever be obtained. The best fitting configuration in k dimensions corresponds to the minimum STRESS, that is

$$S_k = \min_{\mathbf{X}^*} S_1(\mathbf{X}^*).$$

The above procedure was developed by Shepard (1962a, 1962b) and Kruskal (1964) and is therefore known as Kruskal–Shepard's non-metric scaling. Other STRESS formulas that are also used are

$$S_2(\mathbf{X}^*) = \left\{ \sum_{i=1}^{m} (d_i^* - \hat{d}_i)^2 \Big/ \sum_{i=1}^{m} (d_i^* - \bar{d}^*)^2 \right\}^{1/2}$$

$$SS_1(\mathbf{X}^*) = \left\{ \sum_{i=1}^{m} [(d_i^*)^2 - (\hat{d}_i)^2]^2 \Big/ \sum_{i=1}^{m} (d_i^*)^4 \right\}^{1/2}$$

$$SS_2(\mathbf{X}^*) = \left\{ \sum_{i=1}^{m} [(d_i^*)^2 - (\hat{d}_i)^2]^2 \Big/ \sum_{i=1}^{m} [(d_i^*)^2 - (\bar{d}^*)^2]^2 \right\}^{1/2}.$$

$S_1(\mathbf{X}^*)$ and $S_2(\mathbf{X}^*)$ are called STRESS formulas number 1 and 2 respectively. $SS_1(\mathbf{X}^*)$ and $SS_2(\mathbf{X}^*)$ are called S-STRESS formulas number 1 and 2. The last two formulas differ from the first two only in that they are defined in terms of the squared distances. By using different STRESS-formulas and different iterative methods to find the configuration which lead to S_k a wide variety of different techniques are obtained within the Kruskal–Shepard framework.

The goodness of fit is judged graphically by means of the Shepard diagram which is a scatter plot of the m different observed distances d_{ij} (horizontal) against the m different "recovered" distances d_{ij}^* (vertical). If this diagram shows a monotonic nature the scaling is worthwhile. A small STRESS-value is associated with a Shepard diagram which shows only small deviations from monotonicity.

Another measure that could be used to evaluate the fit is the coefficient of determination R^2 of the linear relationship between the recovered distances d_i^* and the disparities $\hat{d}_i$. Large values of R^2 will correspond to small STRESS values.

The choice of k is obviously, as in metric scaling, vitally important. It is customary to determine the best $\mathbf{X}^*$ configuration for different values of k. A scree graph of the minimum STRESS against k gives an indication of a suitable choice of k. In the following representation $k = 2$ appears to be a meaningful choice since $k \geq 3$ does not offer significantly more information than $k = 2$.

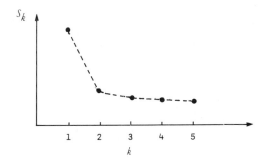

In order to illustrate non-metric scaling two examples will be given. In both examples the SAS PROC ALSCAL program is used. [See SUGI Supplemental Library User's Guide (1983).]

EXAMPLE 1. Non-metric scaling of Ability Test data, variables X_1 to X_{18} (see Table 1.1).

Suppose the 28 × 18 data matrix is converted to a 18 dimensional symmetric distance or dissimilarity matrix using the Pearson distance measure (see Chapter 5). This matrix is shown in Table 6.3.

Table 6.3. Pearson's distance matrix ($\times\,100$) for the data of Table 1.1 (X_1 to X_{18})

	1	2	3	4	5	6	7	8	9	10	11	12	13	14	15	16	17	18	19	20	21	22	23	24	25	26	27	28
1	00																											
2	409	00																										
3	414	512	00																									
4	748	768	501	00																								
5	432	513	699	987	00																							
6	453	510	553	810	688	00																						
7	342	327	396	735	520	448	00																					
8	510	587	323	580	784	598	525	00																				
9	549	587	367	429	811	562	496	401	00																			
10	396	519	364	727	580	391	365	514	514	00																		
11	761	827	508	499	1041	786	778	602	512	747	00																	
12	251	460	413	744	446	460	372	519	566	450	810	00																
13	665	714	418	386	933	666	643	446	327	628	464	665	00															
14	527	681	440	542	893	624	638	579	521	642	580	590	533	00														
15	498	582	350	575	784	558	537	514	408	594	441	521	500	425	00													
16	957	1021	707	530	1248	935	940	787	632	990	464	938	532	696	658	00												
17	603	684	501	627	865	615	635	517	528	652	632	602	509	482	489	721	00											
18	537	636	416	510	855	674	585	477	444	584	532	573	486	427	435	653	422	00										
19	744	826	542	411	1078	826	803	588	529	798	394	788	419	423	518	439	516	407	00									
20	556	701	455	461	880	755	671	512	477	639	568	633	415	377	536	659	567	447	360	00								
21	447	403	445	766	488	548	381	433	555	368	842	397	627	736	618	1024	614	675	854	700	00							
22	363	422	455	747	526	554	408	458	587	356	810	446	683	610	634	1028	550	599	774	611	356	00						
23	572	450	557	789	689	481	440	559	551	525	854	573	610	765	709	1023	723	755	872	747	395	494	00					
24	444	477	466	718	625	520	393	620	469	608	668	433	615	614	402	821	588	542	712	660	549	585	573	00				
25	382	403	534	867	480	451	413	590	622	465	894	421	715	703	648	1112	727	763	913	717	359	403	343	574	00			
26	503	598	397	660	721	546	540	354	511	412	610	511	533	561	501	804	402	514	630	570	443	412	623	610	596	00		
27	437	470	477	715	619	505	422	482	480	454	766	517	614	632	581	928	642	687	776	605	422	409	482	604	453	490	00	
28	513	528	348	455	758	568	503	366	291	480	535	514	413	479	428	707	486	447	511	446	490	467	523	457	554	404	510	00

Input: SAS PROC ALSCAL program for Kruskal–Shephard's non-metric scaling of the ability test dissimilarity matrix

```
DATA    DISTABIL ;
INPUT   (DIST1 - DIST28 ) ( 14*5.2/14*5.2 ) ;
CARDS ;
 0.0

 4.09

 4.14 5.12

 7.48 7.68 5.01

      .

      .

      .

 4.37 4.70 4.77 7.15 6.19 5.05 4.22 4.82 4.80 4.54 7.66 5.17 6.14 6.32
 5.81 9.28 6.42 6.87 7.76 6.05 4.22 4.09 4.82 6.04 4.53 4.90
 5.13 5.28 3.48 4.55 7.58 5.68 5.03 3.66 2.91 4.80 5.35 5.14 4.13 4.79
 4.28 7.07 4.86 4.47 5.11 4.46 4.90 4.67 5.23 4.57 5.54 4.04 5.10
 ;
PROC    ALSCAL MODEL=EUCLID  LEVEL=ORDINAL MINDIM=1 MAXDIM=6
        PRINT PLOT ;
TITLE   'Non-metric scaling of 28 Pupils according to ability tests' ;
```

The ALSCAL program uses only the data appearing in the lower triangle of the symmetric data matrix. Therefore, the reading of blanks as zeroes in the upper triangle of the matrix does not affect the analysis.

Output:

Although the above program performs subsequent 6-, 5-, 4-, 3-, 2- and 1-dimensional scalings (MAXDIM and MINDIM options) with pairwise plots of all the axes as well as the plots of d_i^* against $\hat{d}_i$ and of d_i^* against d_i only some of the output for the two- and one-dimensional scaling is shown.

(i) *Two-dimensional scaling*

Iteration history, S_2 and R^2:

```
ITERATION HISTORY FOR THE 2 DIMENSIONAL SOLUTION (IN SQUARED DISTANCES)
              YOUNGS S-STRESS FORMULA 1 IS USED.

       ITERATION        S-STRESS        IMPROVEMENT

           1             0.16701
           2             0.13244         0.03457
           3             0.12949         0.00295
           4             0.12881         0.00068

                   ITERATIONS STOPPED BECAUSE
            S-STRESS IMPROVEMENT LESS THAN 0.001000


          STRESS AND SQUARED CORRELATION (RSQ) IN DISTANCES

RSQ VALUES ARE THE PROPORTION OF VARIANCE OF THE SCALED DATA (DISPARITIES) IN THE PARTITION
  (ROW, MATRIX, OR ENTIRE DATA) WHICH IS ACCOUNTED FOR BY THEIR  CORRESPONDING DISTANCES.

            STRESS VALUES ARE KRUSKAL'S STRESS FORMULA 1.

                STRESS = 0.138      RSQ = 0.920
```

Optimal X^* configuration.

| | | DIMENSION | |
STIMULUS NUMBER	PLOT SYMBOL	1	2
1	1	0.8789	-0.3444
2	2	1.3408	-0.3394
3	3	-0.1660	0.0791
4	4	-1.7413	0.6045
5	5	2.5903	-0.2773
6	6	1.0662	-0.8433
7	7	1.0480	-0.0611
8	8	-0.1672	0.6879
9	9	-0.5103	0.2250
10	A	0.9478	0.6051
11	B	-2.0121	-0.0105
12	C	0.9965	-0.4323
13	D	-1.1349	0.4334
14	E	-0.9567	-0.7821
15	F	-0.5787	-0.6591
16	G	-3.0706	-0.2249
17	H	-0.6412	-0.9900
18	I	-0.8805	-0.4945
19	J	-2.0645	-0.1281
20	K	-1.1417	0.4785
21	L	1.2479	0.4075
22	M	1.0592	0.3895
23	N	1.2702	1.0075
24	O	0.4134	-0.9948
25	P	1.6628	0.1911
26	Q	-0.0154	0.5905
27	R	0.8307	0.7244
28	S	-0.2715	0.1580

Scatter plot of pupils (Pupils 5 and 16 are not represented since their co-ordinates for the first dimension fall outside the range).

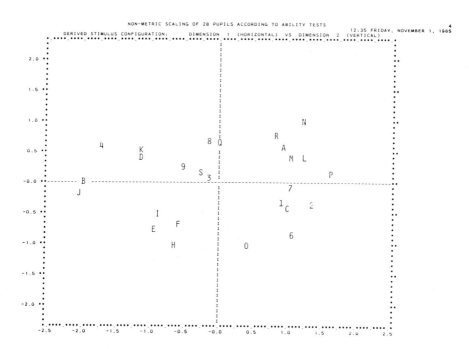

Shepard diagram of monotonic regression:

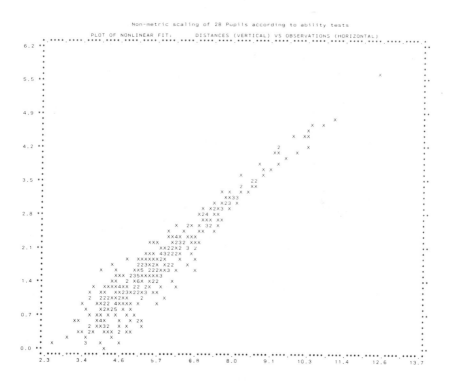

(ii) *One-dimensional scaling*

Iteration history, S_1 and R^2.

```
ITERATION HISTORY FOR THE 1 DIMENSIONAL SOLUTION (IN SQUARED DISTANCES)
             YOUNGS S-STRESS FORMULA 1 IS USED.

        ITERATION      S-STRESS       IMPROVEMENT

            1           0.19217
            2           0.16824        0.02393
            3           0.16702        0.00122
            4           0.16677        0.00025

                  ITERATIONS STOPPED BECAUSE
          S-STRESS IMPROVEMENT LESS THAN 0.001000


        STRESS AND SQUARED CORRELATION (RSQ) IN DISTANCES

RSQ VALUES ARE THE PROPORTION OF VARIANCE OF THE SCALED DATA (DISPARITIES) IN THE PARTITION
   (ROW, MATRIX, OR ENTIRE DATA) WHICH IS ACCOUNTED FOR BY THEIR  CORRESPONDING DISTANCES.

        STRESS VALUES ARE KRUSKAL'S STRESS FORMULA 1.

                 STRESS = 0.217      RSQ = 0.868
```

Optimal **X*** configuration.

| | | DIMENSION |
STIMULUS NUMBER	PLOT SYMBOL	1
1	1	-0.6245
2	2	-1.0091
3	3	0.1423
4	4	1.3470
5	5	-1.9883
6	6	-0.8829
7	7	-0.7564
8	8	0.1335
9	9	0.3409
10	A	-0.7351
11	B	1.5187
12	C	-0.7318
13	D	0.8218
14	E	0.7787
15	F	0.4593
16	G	2.3579
17	H	0.6096
18	I	0.6729
19	J	1.5756
20	K	0.8609
21	L	-0.9603
22	M	-0.7849
23	N	-1.0897
24	O	-0.3583
25	P	-1.2335
26	Q	0.0122
27	R	-0.6541
28	S	0.1774

Ordering of pupils.

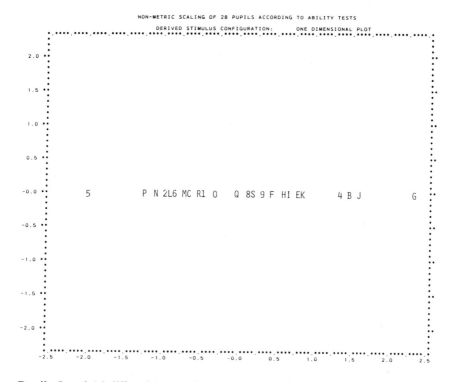

Pupils 5 and 16 differ the most in respect of achievement. The fact that the one-dimensional scaling awards Pupil 5 a negative value and Pupil 16 a positive value has no significance since only the ordering is of importance. The scores obtained by multiplying the coordinates by -1 also produces a correct scaling.

(iii) *Scree graph of minimum STRESS* (plotted by using the minimum STRESS-values (Kruskal's STRESS formula 1) that appear in the output)

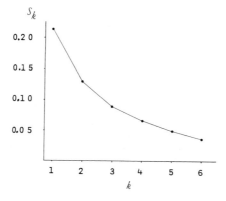

This graph shows that a two-dimensional scaling is not quite satisfactory in describing the 18 dimensional data set.

The one-dimensional ordening and two-dimensional scatterplot of the 28 pupils will be used in Chapter 11 to investigate possible location and scale differences between boys and girls in respect of the 18 ability tests.

EXAMPLE 2. Non-metric scaling of occupation data (see Section 5).

Input: SAS PROC ALSCAL programme for Kruskal–Shepard's non-metric scaling of occupation dissimilarity matrix

```
DATA    OCCUPAT ;
INPUT   (PREACHER SURGEON TEACHER JOURNAL POLICE
        PLUMBER FARMER ACTOR CLERK ) ( 9*5.2 ) ;
CARDS ;
 0.00
 3.06 0.00
 2.14 3.95 0.00
 3.21 3.95 3.03 0.00
 3.51 4.17 2.82 3.33 0.00
 4.40 3.77 3.86 4.14 3.60 0.00
 3.64 3.69 3.47 3.90 3.56 2.53 0.00
 3.12 4.13 2.72 2.58 3.68 4.17 4.10 0.00
 3.73 4.05 3.31 3.46 3.29 3.92 3.59 3.72 0.00
 ;
PROC    ALSCAL MODEL=EUCLID  LEVEL=ORDINAL MINDIM=1 MAXDIM=4
        PRINT PLOT ;
TITLE   'Non-metric scaling of 9 occupations' ;
```

Since the 9×9 dissimilarity matrix consists of only 36 non-diagonal elements the highest order configuration that can be found by the Kruskal–Shepard method is four-dimensional.

Output:

(i) *Four-dimensional scaling*

Iteration history, S_4 and R^2.

```
ITERATION HISTORY FOR THE 4 DIMENSIONAL SOLUTION (IN SQUARED DISTANCES)
              YOUNGS S-STRESS FORMULA 1 IS USED.

        ITERATION      S-STRESS        IMPROVEMENT
            1           0.07349
            2           0.04740         0.02609
            3           0.04126         0.00614
            4           0.03752         0.00375
            5           0.03590         0.00162
            6           0.03497         0.00093

             ITERATIONS STOPPED BECAUSE
        S-STRESS IMPROVEMENT LESS THAN 0.001000


        STRESS AND SQUARED CORRELATION (RSQ) IN DISTANCES

RSQ VALUES ARE THE PROPORTION OF VARIANCE OF THE SCALED DATA (DISPARITIES) IN THE PARTITION
(ROW, MATRIX, OR ENTIRE DATA) WHICH IS ACCOUNTED FOR BY THEIR  CORRESPONDING DISTANCES.

        STRESS VALUES ARE KRUSKAL'S STRESS FORMULA 1.

            STRESS = 0.031      RSQ = 0.987
```

Optimal **X*** configuration.

Non-metric scaling of 9 occupations

CONFIGURATION DERIVED IN 4 DIMENSIONS

STIMULUS COORDINATES

STIMULUS NUMBER	PLOT SYMBOL	DIMENSION 1	2	3	4
1	1	-0.9538	-1.1972	0.2562	-0.8500
2	2	0.7024	-2.3346	0.3864	0.6584
3	3	-0.9104	0.2694	-0.0746	-0.6939
4	4	-1.3779	0.3048	-0.5310	0.7926
5	5	-0.1027	1.2590	0.7471	-1.0077
6	6	2.3918	0.4957	-0.6947	0.0907
7	7	1.8002	0.1375	-0.3038	-0.5269
8	8	-1.5855	0.1166	-1.2685	0.3332
9	9	0.0360	0.9489	1.4827	1.2036

(iii) *One-dimensional scaling*

Iteration history, S_1 and R^2.

ITERATION HISTORY FOR THE 2 DIMENSIONAL SOLUTION (IN SQUARED DISTANCES)
YOUNGS S-STRESS FORMULA 1 IS USED.

ITERATION	S-STRESS	IMPROVEMENT
1	0.19072	
2	0.13424	0.05647
3	0.12999	0.00425
4	0.12910	0.00089

ITERATIONS STOPPED BECAUSE
S-STRESS IMPROVEMENT LESS THAN 0.001000

STRESS AND SQUARED CORRELATION (RSQ) IN DISTANCES

RSQ VALUES ARE THE PROPORTION OF VARIANCE OF THE SCALED DATA (DISPARITIES) IN THE PARTITION
(ROW, MATRIX, OR ENTIRE DATA) WHICH IS ACCOUNTED FOR BY THEIR CORRESPONDING DISTANCES.

STRESS VALUES ARE KRUSKAL'S STRESS FORMULA 1.

STRESS = 0.117   RSQ = 0.916

Optimal **X*** configuration.

STIMULUS NUMBER	PLOT SYMBOL	DIMENSION 1	2
1	1	-0.6922	-0.9420
2	2	0.6554	-1.8814
3	3	-0.6003	0.3399
4	4	-1.1339	0.0983
5	5	-0.0262	1.2078
6	6	1.9995	0.3811
7	7	1.4221	-0.1657
8	8	-1.4295	-0.1089
9	9	-0.1950	1.0710

Scatter plot of occupations.

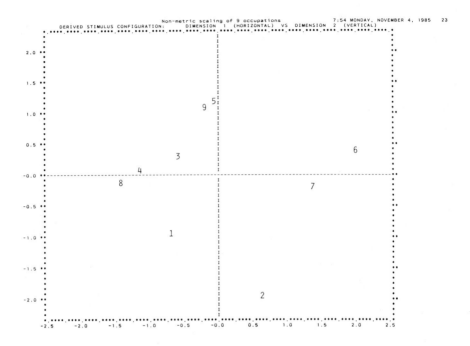

(iii) *One-dimensional scaling*

Iteration history, S_1 and R^2.

```
YOUNGS S-STRESS FORMULA 1 IS USED.

ITERATION        S-STRESS            IMPROVEMENT

    1             0.49767
    2             0.41017            0.08750
    3             0.37566            0.03451
    4             0.34836            0.02730
    5             0.33872            0.00963
    6             0.33484            0.00389
    7             0.33330            0.00154
    8             0.33264            0.00065

          ITERATIONS STOPPED BECAUSE
    S-STRESS IMPROVEMENT LESS THAN 0.001000


      STRESS AND SQUARED CORRELATION (RSQ) IN DISTANCES

RSQ VALUES ARE THE PROPORTION OF VARIANCE OF THE SCALED DATA (DISPARITIES) IN THE PARTITION
    (ROW, MATRIX, OR ENTIRE DATA) WHICH IS ACCOUNTED FOR BY THEIR  CORRESPONDING DISTANCES.

      STRESS VALUES ARE KRUSKAL'S STRESS FORMULA 1.

          STRESS = 0.312      RSQ = 0.729
```

Optimal **X*** configuration.

| | | DIMENSION |
STIMULUS NUMBER	PLOT SYMBOL	1
1	1	0.2989
2	2	-1.5958
3	3	0.6901
4	4	1.0208
5	5	0.3384
6	6	-1.5048
7	7	-0.9591
8	8	1.0587
9	9	0.6527

Ordering of occupations.

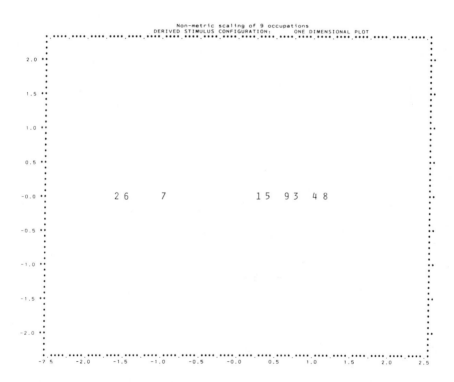

The scatter plots of the two-dimensional metric scaling (see Section 5) and non-metric scaling of the occupation dissimilarity matrix correspond well but the one-dimensional scalings differ to a large extent. If the scatter plot is used for an informal cluster analysis, it is recommended by Gower and Ross (1969) to first obtain a minimal spanning tree by using the observed dissimilarity matrix. The minimal spanning tree of the occupation dissimilarity matrix is

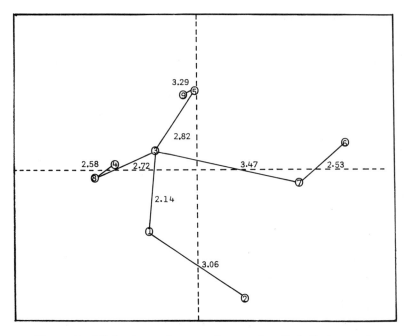

Figure 6.2. Minimal spanning tree of occupation dissimilarity matrix.

shown in Figure 6.2 on the scatter plot of the two-dimensional non-metric scaling which was obtained in the output of Example 2.

The lengths of the edges (line segments) between the nodes are the distances as observed in the dissimilarity matrix and do not correspond to the geometric distances on the two-dimensional plot. A single linkage cluster analysis (see Chapter 5) dividing the occupations in say 3 groups gives the following results.

Group 1: Plumber, farmer
Group 2: Surgeon, teacher, journalist, policeman, actor, preacher
Group 3: Bank clerk

7. Three-Way Multidimensional Scaling (INDSCAL)

INDSCAL ("Individual Differences Multidimensional Scaling") is the name of a method for performing a three-way multidimensional scaling. The INDSCAL model was originally formulated [(Carroll and Chang (1970)] with a view to taking into account individual differences in the scaling of variables (stimuli). A symmetrical dissimilarity matrix is calculated for each individual (or individual group) on the basis of the latter's perception of the stimuli. These matrices constitute the input of an INDSCAL computer program.

The INDSCAL model presupposes a set of r dimensions (factors) subjacent to the way in which the individual observes dissimilarities between the stimuli. It is assumed that the dissimilarity judgments (squared) of each individual or case are related linearly to a weighted Euclidean distance. The weighted Euclidean distance between variables j and k for case i is given by

$$d_{jk}^{(i)} = \sqrt{\sum_{t=1}^{r} W_{it}(Y_{jt} - Y_{kt})^2}$$

where Y_{jt} indicates the coordinate of variable j in respect of dimension (factor) t and W_{it} is a weight that indicates the importance of dimension t for case i. A zero weight indicates that the particular dimension (or factor) has no relevancy for a specific individual.

Negative weights should not occur and where they do, this indicates that INDSCAL is not an appropriate model for describing the data.

A two-dimensional scaling usually offers a satisfactory approximation to the model. The measure for evaluating the fit is

$$\bar{R} = \sqrt{R_1^2 + R_2^2 + \cdots + R_n^2}$$

where R_i^2, $i = 1, 2, \ldots, n$ is the multiple correlation coefficient in respect of individual i and is a measure of the correspondence between the observed and fitted dissimilarities of individual i. Two sets of co-ordinates are given as part of the INDSCAL output. The first set is associated with the plot of the joint stimuli space and the second with the plot of the individuals. In the two-dimensional case the output appears as follows:

Stimulus no.	Stimulus co-ordinates	
	Dimension 1	Dimension 2
1	Y_{11}	Y_{12}
2	Y_{21}	Y_{22}
$\vdots$	$\vdots$	$\vdots$
p	Y_{p1}	Y_{p2}

Individual no.	Weights for individuals	
	Dimension 1	Dimension 2
1	W_{11}	W_{12}
2	W_{21}	W_{22}
$\vdots$	$\vdots$	$\vdots$
n	W_{n1}	W_{n2}

A so-called perceptual space can be constructed for each individual by using the relation

$$V_{jt}^{(i)} = \sqrt{W_{ij}}\, Y_{jt}$$

where $V_{jt}^{(i)}$ indicates the co-ordinate of variable j for dimension t in respect of individual i. The perceptual space can therefore be considered as the stimulus space for a particular individual.

Assume the data consists of three stimuli as observed by two individuals and that the computer output is as follows:

A. Stimulus Co-Ordinates

	Dimension 1	Dimension 2
Stimulus 1	−0.5	−1.6
Stimulus 2	−1.3	−1.2
Stimulus 3	1.6	1.8

B. Weights for Individuals

	Dimension 1	Dimension 2
Individual 1	0.6	0.4
Individual 2	0.8	0.2

On the basis of this output a perceptual space can, for instance, be constructed for Individual 1:

C. Perceptual Space: Individual 1

	Dimension 1	Dimension 2
Stimulus 1	$-0.5\sqrt{0.6} = -0.39$	$-1.6\sqrt{0.4} = -1.01$
Stimulus 2	$-1.3\sqrt{0.6} = -1.01$	$-1.2\sqrt{0.4} = -0.76$
Stimulus 3	$1.6\sqrt{0.6} = 1.24$	$1.8\sqrt{0.4} = 1.14$

In the figure below (a) indicates the joint stimulus space, (b) the perceptual space for Individual 1 and (c) a scaling of the cases.

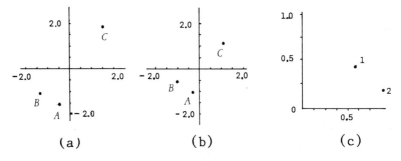

(a)   (b)   (c)

Key: A = Stimulus 1, B = stimulus 2, C = stimulus 3, 1 = individual 1, and 2 = individual 2. Horizontal axis = dimension 1 and vertical axis = dimension 2.

The plotting of perceptual spaces is particularly relevant in longitudinal studies where structural changes in variables over time, are studied. In this case each successive study is regarded as an individual (or new situation) and it is customary to superimpose the perceptual spaces on one another for purposes of comparison.

A vital characteristic of INDSCAL is the unique determination of the dimensions, in that rotations or transformations thereof produce a non-optimal solution. In practice this means that the INDSCAL dimensions (factors) frequently correspond to basic psychological, perceptual and conceptual processes, which may vary from individual to individual. Such differences are illustrated by means of the scaling of the individuals.

For illustrative purposes consider a survey in which test battery scores in respect of eight interest fields were obtained. According to Cudeck (1982) the following structural relationship exists between these fields.

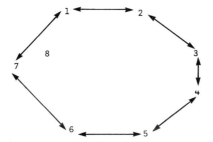

KEY:

1. Health
2. Science
3. Technology
4. Commerce
5. Business transactions
6. Business contact
7. Social
8. Arts

There is a slight dissimilarity (large similarity) between adjacent interest fields as opposed to a great dissimilarity between fields that are situated far apart on the diagram. According to Cudeck a mild dissimilarity exists between arts and the other fields.

The INDSCAL model has been used to indicate whether the above structural relationship is retained irrespective of the group from which the sample is drawn. Four population groups, namely Coloureds, Asians, Afrikaans-speaking and English-speaking Whites, were involved in the investigation. A correlation matrix was calculated for each subgroup on the basis of the scores obtained in an interest field test battery. For illustrative purposes the analysis has been limited to the data on the female respondents.

Since a correlation matrix amounts to a similarity matrix, correlations were converted into dissimilarity measures by means of the transformation $d_{ij} = 100(1 - |r_{ij}|)$. The eight interest fields are the stimuli whereas the four subgroups are regarded as the individuals. The INDSCAL model was fitted by means of the SAS PROC ALSCAL program. The computer input and a part of the computer output are as follows:

Input: SAS PROC ALSCAL program for a three-way scaling according to the INDSCAL method

```
DATA    INTFIELD ;
INPUT   (HEALTH SCIENCE TECHNOL COMMERCE BUSTRANS
        BUSCONT SOCIAL ARTS )( 8*9.5 ) ;
ARRAY   X HEALTH--ARTS ;
        DO OVER X ;
        X=1.0-ABS(X) ;
        X=100.0*X ;
        END ;
* CORRELATION MATRICES ARE IN THE FOLLOWING ORDER: COLOURED FEMALES
  ASIATIC FEMALES, AFRIKAANS-SPEAKING WHITE FEMALES AND ENGLISH-
  SPEAKING WHITE FEMALES ;
CARDS ;
 1.00000
 0.71319  1.00000
 0.71819  0.67006  1.00000
 0.60224  0.55225  0.88149  1.00000
 0.54938  0.59745  0.55338  0.54814  1.00000
 0.62605  0.40918  0.69233  0.70859  0.75560  1.00000
 0.67404  0.54020  0.60500  0.56063  0.66027  0.75878  1.00000
 0.63009  0.58478  0.75589  0.75833  0.48674  0.63634  0.59487  1.00000
 1.00000
 0.79829  1.00000
 0.53671  0.72406  1.00000
 0.33640  0.52050  0.83891  1.00000
 0.34558  0.42666  0.65783  0.69398  1.00000
 0.33472  0.44947  0.62194  0.66754  0.78258  1.00000
 0.52485  0.64500  0.59664  0.49671  0.54004  0.65807  1.00000
 0.31997  0.52735  0.78988  0.76511  0.63615  0.71514  0.58583  1.00000
 1.00000
 0.76749  1.00000
 0.55145  0.73907  1.00000
 0.47440  0.60871  0.84780  1.00000
 0.23205  0.23120  0.49337  0.53444  1.00000
 0.37226  0.35827  0.61826  0.60056  0.73876  1.00000
 0.59063  0.52138  0.54653  0.49477  0.47000  0.73611  1.00000
 0.47765  0.59353  0.71903  0.68568  0.43421  0.64341  0.62598  1.00000
 1.00000
 0.71998  1.00000
 0.52599  0.69391  1.00000
 0.45385  0.51044  0.76966  1.00000
 0.17492  0.16465  0.44785  0.40451  1.00000
 0.30719  0.24753  0.48993  0.38222  0.77109  1.00000
 0.54337  0.40684  0.46775  0.36558  0.44158  0.67157  1.00000
 0.38472  0.49553  0.66507  0.54652  0.36548  0.48675  0.55457  1.00000
 ;
PROC    ALSCAL MODEL=INDSCAL LEVEL=INTERVAL PRINT PLOT ;
TITLE   'Eight interest areas - females' ;
```

Output: Three-way scaling according to the INDSCAL method

(i) *Stress*

Eight interest areas - females

ITERATION HISTORY FOR THE 2 DIMENSIONAL SOLUTION (IN SQUARED DISTANCES)
YOUNGS S-STRESS FORMULA 1 IS USED.

ITERATION	S-STRESS	IMPROVEMENT
0	0.27448	
1	0.27411	
2	0.25550	0.01861
3	0.25422	0.00128
4	0.25400	0.00023

ITERATIONS STOPPED BECAUSE
S-STRESS IMPROVEMENT LESS THAN 0.001000

(ii) *Goodness of fit $\bar{R}$*

AVERAGED (RMS) OVER MATRICES
STRESS = 0.204   RSQ = 0.747

(iii) *Co-ordinates for plotting the stimulus space*

STIMULUS NUMBER	PLOT SYMBOL	DIMENSION 1	DIMENSION 2
1	1	-1.7816	-0.3337
2	2	-1.1183	-1.1628
3	3	0.2040	-0.7231
4	4	0.9411	-0.7852
5	5	0.9004	1.5139
6	6	0.7384	1.1764
7	7	-0.7437	1.0716
8	8	0.8597	-0.7572

(iv) *Co-ordinates for plotting individuals*

SUBJECT WEIGHTS

SUBJECT NUMBER	PLOT SYMBOL	WEIRD- NESS	DIMENSION 1	DIMENSION 2
1	1	0.3130	0.4169	0.6682
2	2	0.5685	0.8870	0.3001
3	3	0.0843	0.5797	0.6354
4	4	0.2159	0.5149	0.6983

OVERALL IMPORTANCE OF EACH DIMENSION   0.3905   0.3570

(v) *Graphical representation of interest fields*

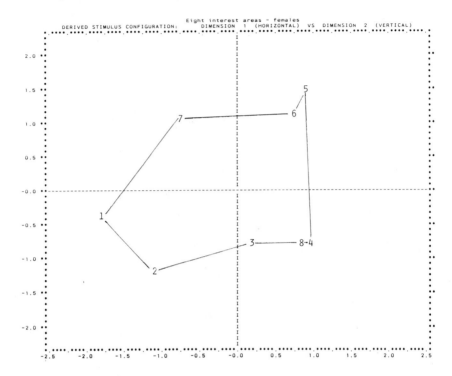

The graphical representation of the interest fields confirms the postulated relationship, with the exception of Field 8, the arts, which is evidently only slightly dissimilar to Fields 3 and 4.

(vi) *Graphical representation of subgroups* (*individuals*)

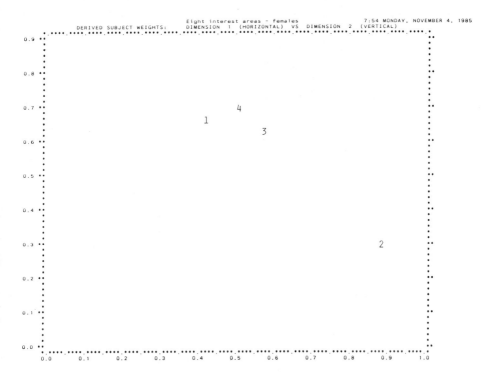

It appears that in relation to the other groups the women from the Asian subgroup have the greatest weight in respect of dimension 1 and the smallest weight in respect of dimension 2. A graphical representation of the perceptual space for Asians will indicate the extent to which this subgroup deviates from the joint stimulus space.

The ALSCAL program has different options, thus enabling the user to fit various types of INDSCAL models to the data. The interested reader is referred to the SUGI Supplemental Library Users' Guide (1983).

8. Guttman's Techniques

In his efforts to find alternatives to classical inference Guttman (1977) has developed various techniques for exploratory data analysis. In this context HUDAP (Hebrew University Data Analysis Package) is a comprehensive program package which incorporates these techniques. In this section two procedures in this program package will be introduced and discussed briefly. In the next two sections the procedures will then be illustrated by means of practical applications.

The program WSSA (Weighted smallest space analysis) is intended as an aid to applications in facet theory. Facet theory [see Canter (1985)] is an approach to the search for lawfulness in the social sciences which was devised by Louis Guttman and subsequently developed by him and his colleagues. This theory offers a set of principles as a guide to research design and makes use of non-metric scaling to analyze the data. In facet theory use is made of the interpretation of configurations of variables in multidimensional space as a means of establishing a framework within which behavioural theories, for example, can be constructed.

The key to facet theory is the establishment of a mapping sentence. A mapping sentence consists of three types of facets and may be represented symbolically as follows:

$$X \ ABC\ldots N \to R.$$

In this representation the first type of facet, denoted by X, designates the populaton of respondents being researched. Facets of the second kind classify the content of the variables and are denoted by $ABC\ldots N$. $X \ ABC\ldots N$ together define the domain of the mapping sentence. The third kind of facet is the range and the arrow denotes the mapping of the domain onto the set R of possible responses. Each respondent (x) has one and oniy one response in R for each item (e.g. question) classified by the elements of the content facets $ABC\ldots N$.

As an illustration consider the following eight-item questionaire:

Some people will agree (or disagree) about the existence of sufficient opportunities with regard to the following issues. Please indicate your point of view.

1. To further one's own studies   agree ⬚⬚⬚⬚⬚ disagree

2. To improve one's own income   ⬚⬚⬚⬚⬚

3. To meet interesting people   ⬚⬚⬚⬚⬚

4. For others to further their studies   ⬚⬚⬚⬚⬚

5. For others to improve their income   ⬚⬚⬚⬚⬚

6. For others to meet interesting people   ⬚⬚⬚⬚⬚
7. To find other employment if I lose my
   present job   ⬚⬚⬚⬚⬚
8. For others to find employment if they
   lose their jobs   ⬚⬚⬚⬚⬚

A mapping sentence for the above eight questions is

$$X \ AB \to R.$$

A less abstract formulation of the mapping sentence is:

The level of agreement of respondent (x) regarding the existence of opportunities for

$$
\begin{array}{cc}
A & B \\
\begin{array}{l}
\text{1. himself} \\
\text{2. others}
\end{array} \Big\} \text{ with regard to} &
\left.\begin{array}{l}
\text{1. social} \\
\text{2. economics} \\
\text{3. security}
\end{array}\right\} \text{ is } \left(\begin{array}{l}\text{high}\\ \text{to}\\ \text{low}\end{array}\right).
\end{array}
$$

Each item in the above questionaire can possibly be described in terms of the elements $a_i b_j$ of the content facets A and B as follows.

Item no.	Elements of content facets
1	$a_1 b_3$
2	$a_1 b_2$
3	$a_1 b_1$
4	$a_2 b_3$
5	$a_2 b_2$
6	$a_2 b_1$
7	$a_1 b_3$
8	$a_2 b_3$

WSSA performs a multidimensional scaling of items. Suppose a three-dimensional scaling of the eight items is carried out on a matrix of association coefficients between these items and that the following is a plot of the first two dimensions:

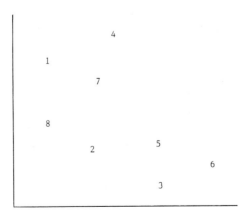

Plot of axis 2 versus axis 1

WSSA also produces facet diagrams, each facet diagram being a replacement of the variables by the elements of the facet under consideration. A facet diagram for facet B is given below.

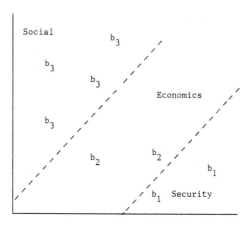

Facet B diagram

In the above example the subspace of the correlation (similarity) matrix of the eight variables is partitioned into three regions. This correspondence between regions of the subspace and the elements of facets is the basis for establishing so-called regional hypotheses (laws).

The computer program POSAC (Partial order scalogram analysis) may be used to analyse data [Shye (1978)] provided that: (i) A substantive rationale guided the selection of variables to be processed together and (ii) the range of each variable is ordered and the order is uniform in its direction and general meaning for all variables included.

The data matrix obtained from such variables (items) is called a scalogram. POSAC gives an optimal two-dimensional representation of the minimal-space diagram for the different profiles (rows) of the observed scalogram. This representation depends on the partial order that exists between the various profiles. The following is a typical representation of a scalogram in two-dimensional space. The respondent's sex is given in brackets after each profile:

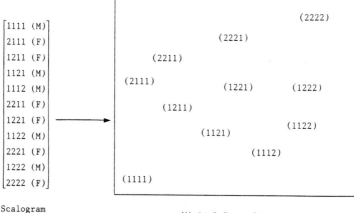

Scalogram
for items   Minimal Space Diagram

POSAC also produces two-dimensional item diagrams. The diagram for item number 1 is given below. This diagram corresponds exactly with the two-dimensional configuration given earlier but only the elements of item 1 are shown on the plot.

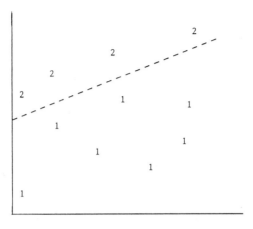

Item 1 diagram

The same graphical print-out may also be used to relate the partial order scalogram with background traits such as sex, language, marital status etc. Consider, for example, the background trait diagram for sex:

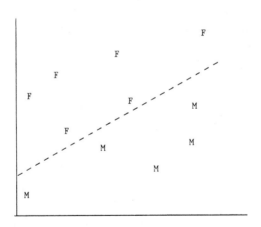

Background trait diagram for sex

If the multivariate distribution is adequatley represented in a two-dimensional partly ordered space, then an optimal discriminant function based on an external trait should be at most two-dimensional. In the above example it turns out that the discriminant function with respect to the external trait sex is a line (i.e., a one-dimensional function).

An important aspect of the contributions of Guttman and his co-workers is the fact that empirical research projects must be designed very carefully to ensure the optimal use of the facet theory approach and the analysis of ordinal data by means of POSAC. This implies that logical thinking and careful planning must take place before such matters as the finalization of questionnaires and the actual collection of data.

9. Facet Theory

Guttman defines a theory as an "hypothesis of a correspondence between a definitional system for a universe of observations and an aspect of the empirical structure of those observations, together with a rationale for such an hypothesis" [Guttman in Gratch (1973)]. This definition implies that the researcher should carefully consider:

(1) the experimental design on which his observations were based, and
(2) the empirical structure of those observations.

In Guttman's terminology First Laws of Intercorrelations are concerned only with the sign of correlations among certain kinds of variables (for example attitude, intelligence and involvement). Regional hypotheses are based on relative sizes of correlations and are called Second Laws of Intercorrelations. These hypotheses are associated with the geometry of Smallest Space Analysis (SSA) which utilizes the computer program WSSA. This program is based on the technique of non-metric multidimensional scaling [Guttman (1968)]. Each variable (item) is treated as a point in an Euclidean space in such a way that, the higher the correlation between two variables, the closer they are in this Euclidean space. The space of smallest dimensionality, which makes an inverse relationship between the observed correlations and the geometrical distances possible, is used.

Laws concerning the size of correlations are based on the definitional system of the content of the items to be studied. For a given problem it is important to have the content design specify both the similarities and differences between the items to be analyzed. A convenient way for doing this is by use of a mapping sentence (see Section 8).

According to Levy (1985) fruitful strategies are made possible by use of mapping sentences since the latter lend themselves easily to correction, deletion and extension. Regional hypotheses relate the several roles that the content facets of the variables can play in partitioning the SSA space of the empirical correlation matrix of those variables. The roles of facets and the possible partitioning of space is shown in Figure 6.3.

When studying the association between pairs of variables perfect trend

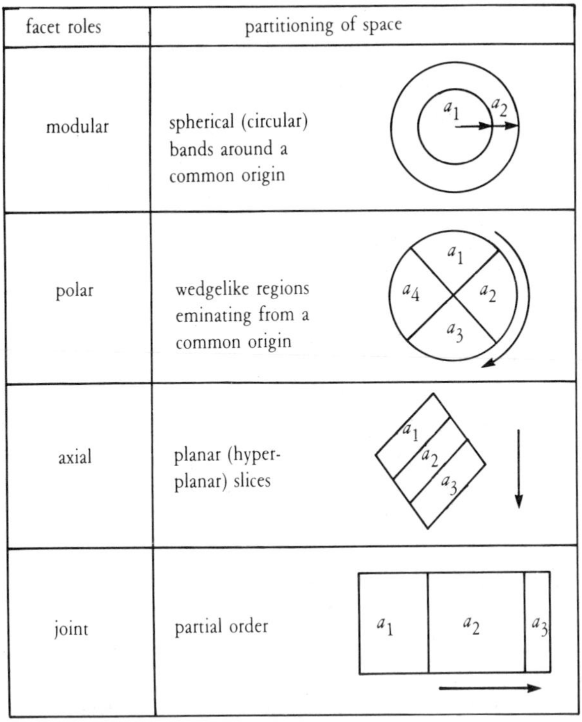

facet roles	partitioning of space	
modular	spherical (circular) bands around a common origin	
polar	wedgelike regions eminating from a common origin	
axial	planar (hyper-planar) slices	
joint	partial order	

Figure 6.3. Roles of facets and their spatial partitions.

rarely occurs in empirical problems. Efficacy of fit of the trend to the data can be assessed in many ways. Popular types of coefficients for this purpose are correlation coefficients. Guttman's coefficient of (weak) monotonicity is an example of such an efficacy coefficient. A monotone trend of Y on X implies that as X increases, any change in Y is one-directional with the possibility that Y can occasionally remain constant. Guttman's coefficient of monotonicity applies equally well to items with discrete attitudinal response categories (such as "strongly agree," "agree," "disagree," "strongly disagree") and to numerically valued continuous variables. The coefficient is defined as

$$\mu_2 = \frac{\sum_i \sum_j (X_i - X_j)(Y_i - Y_j)}{\sum_i \sum_j |X_i - X_j| |Y_i - Y_j|},$$

which varies in absolute value between 0 and 1, where 1 expresses perfect fit.

In facet research, the matrix of intercorrelations is often derived from the coefficients of monotonicity. These μ_2-values are in general greater in absolute value than Pearson's product moment correlation coefficients and are usually applicable to a larger class of variable types.

The Cylinder of Socio-Political Change

In South Africa various institutions regularly undertake studies regarding a wide variety of socio-political matters among various population groups.

To gain more insight in the public views on major changes on the socio-political front, a questionnaire was devised to determine how South Africans feel about certain typical problems in this regard. As an illustration, ten questions from this questionnaire were analysed using the computer program WSSA. The observations were collected during March 1985. The relevant questions are:

Question 1: How do you think the general economic situation in South Africa has changed over the last year?

Question 2: How do you think the general economic situation in South Africa will change over the next year?

Question 3: How do you think the general political situation in South Africa has changed over the last year?

Question 4: How do you think the general political situation in South Africa will change over the next year?

Question 5: How do you think the general relationship between the Whites and the Coloured population groups has changed over the last year?

Question 6: How do you think the general relationship between the Whites and the Coloured population groups will change over the next year?

Question 7: How do you think the general relationship between the White and the Asian population groups has changed over the last year?

Question 8: How do you think the general relationship between the White and the Asian population groups will change over the next year?

Question 9: How do you think the Government has fared with its plans for effecting political change over the last year?

Question 10: How do you think the Government will fare with its plans for effecting political change during the next year?

Each question had to be answered on a scale ranging from has it (will it): improve, remain the same, deteriorate.

A mapping sentence for observations on Socio-political change is:

The way in which respondent (x) thinks about the

$\begin{Bmatrix} 1. & \text{present} \\ 2. & \text{future} \end{Bmatrix}$ with regard to the issue $\qquad$ $\begin{Bmatrix} 1. & \text{economy} \\ 2. & \text{political} \\ 3. & \text{social} \end{Bmatrix}$ $\begin{matrix} \rightarrow \text{improve} \\ \text{to} \\ \text{deteriorate} \end{matrix}$

$\qquad\quad$ A $\qquad\qquad\qquad\qquad\qquad\qquad\quad$ B

This mapping sentence has two content facets and each of the 10 attitude items

Table 6.4. Monotonicity coefficients (decimal
points omitted) between the socio-political
change items ($n = 591$)

		1	2	3	4	5	6	7	8	9	10
ECONPRES	1	100	43	45	31	38	22	6	16	43	21
ECONFUTR	2	43	100	36	51	37	46	33	45	42	26
POLPRES	3	45	36	100	84	55	65	49	54	72	66
POLFUTR	4	31	51	84	100	59	62	41	49	66	60
COLPRES	5	38	37	55	59	100	75	56	54	51	43
COLFUTR	6	22	46	65	62	75	100	54	72	53	49
ASIPRES	7	6	33	49	41	56	54	100	80	44	26
ASIFUTR	8	16	45	54	49	54	72	80	100	49	28
GOVPRES	9	43	42	72	66	51	53	44	49	100	56
GOVFUTR	10	21	26	66	60	43	49	26	28	56	100

is easily classified by these facets. The monotonicity coefficients between these
items are given in Table 6.4.

The HUDAP (see Section 8) procedures MONCO and WSSA were used
to compute the monotonicity coefficients and to perform a smallest space
analysis.

Input: HUDAP procedures MONCO and WSSA for performing a smallest space
analysis

```
$RUN NAME =  'The cylinder of Socio-Political change ' ;
$ITEM NAMES ARE
              ECONPRES ECONFUTR POLPRES POLFUTR COLPRES
              COLFUTR  ASIPRES ASIFUTR GOVPRES GOVFUTR ;
       MISSINGS ARE ECONPRES TO GOVFUTR (-1) ;
$DATA
              FILE =  'MYFILE1' ;
              FORMAT =  '(10F3.0)' ;
$OUTPUT
              MEMORY ;
```

```
$MONCO        NAMES ARE
              ECONPRES ECONFUTR POLPRES POLFUTR COLPRES
              COLFUTR  ASIPRES ASIFUTR GOVPRES GOVFUTR ;
$WSSA1        NAMES ARE
              ECONPRES ECONFUTR POLPRES POLFUTR COLPRES
              COLFUTR  ASIPRES ASIFUTR GOVPRES GOVFUTR ;
              MAXDIM = 3 ;
              FACETS
                        NFACETS = 2 /
                        PROFILES ARE
                               ECONPRES   ( 1 1 )
                               ECONFUTR   ( 1 2 )
                               POLPRES    ( 2 1 )
                               POLFUTR    ( 2 2 )
                               COLPRES    ( 3 1 )
                               COLFUTR    ( 3 2 )
                               ASIPRES    ( 3 1 )
                               ASIFUTR    ( 3 2 )
                               GOVPRES    ( 2 1 )
                               GOVFUTR    ( 2 2 ) /
                        DIAGRAMS = ( 3 1 1 2 ) ,
                                   ( 3 2 1 3 ) ;
{ DIAGRAMS = (A B C D )  Where A = Dimensionality
                               B = Facet number
                               C = Axis no. I
                               D = Axis no. J   }
$END
```

Facet diagrams are produced by specifying the number of facets, the contents of each facet and the axis numbers in the FACETS paragraph of WSSA. Relevant sections of the computer output are as follows.

Output:

(i) *Co-ordinates for dimensionality* 3

```
RANK IMAGE TRANSFORMATIONS ............. 17
NUMBER OF ITERATIONS ................... 32
COEFFICIENT OF ALIENATION .............. 0.06813
```

SERIAL NUMBER	DISTANCE FROM CENTROID	PLOTTED COORDINATES 1	2	3
1	76.50	0.00	25.69	0.00
2	63.00	55.46	0.00	57.32
3	21.61	44.42	78.72	49.35
4	25.62	41.80	63.62	65.79
5	29.79	65.25	79.23	22.62
6	26.26	75.94	70.65	50.21
7	55.15	100.00	65.05	14.74
8	42.25	94.51	58.13	58.20
9	32.36	24.93	77.60	46.15
10	56.77	23.81	93.42	79.52

(ii) *Space diagram for dimensionality* 3, *Axis* 1 *versus Axis* 2

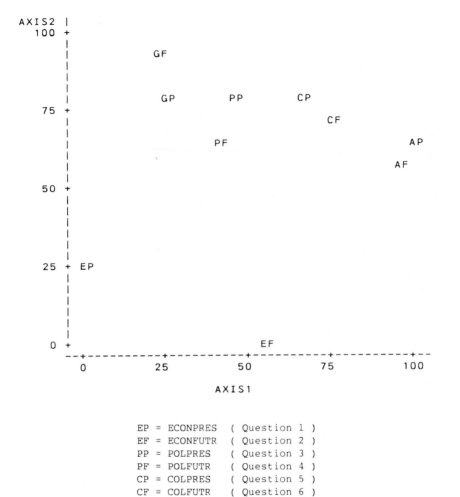

```
EP  =  ECONPRES    ( Question  1 )
EF  =  ECONFUTR    ( Question  2 )
PP  =  POLPRES     ( Question  3 )
PF  =  POLFUTR     ( Question  4 )
CP  =  COLPRES     ( Question  5 )
CF  =  COLFUTR     ( Question  6 )
AP  =  ASIPRES     ( Question  7 )
AF  =  ASIFUTR     ( Question  8 )
GP  =  GOVPRES     ( Question  9 )
GF  =  GOVFUTR     ( Question 10 )
```

(iii) *Facet diagram for Facet B, Axis 1 versus axis 2*

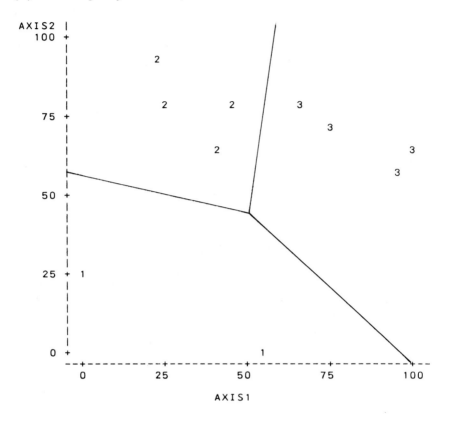

KEY :

1 = Economy

2 = Political

3 = Social

The space is partitioned in wedgelike regions by the elements of facet *B*.

(iv) *Space diagram for dimensionality* 3, *Axis* 1 *versus Axis* 3

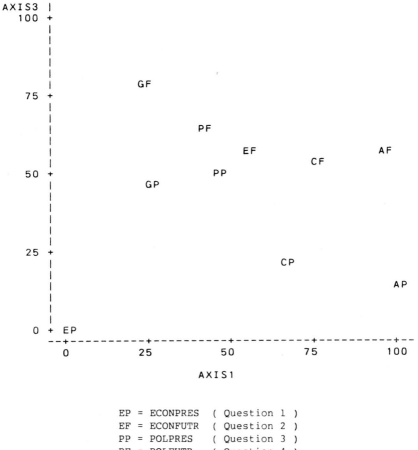

EP	=	ECONPRES	(Question 1)
EF	=	ECONFUTR	(Question 2)
PP	=	POLPRES	(Question 3)
PF	=	POLFUTR	(Question 4)
CP	=	COLPRES	(Question 5)
CF	=	COLFUTR	(Question 6)
AP	=	ASIPRES	(Question 7)
AF	=	ASIFUTR	(Question 8)
GP	=	GOVPRES	(Question 9)
GF	=	GOVFUTR	(Question 10)

(v) *Facet diagram for Facet A, Axis 1 versus Axis 3*

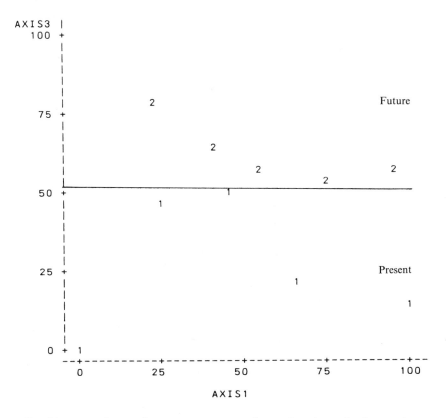

In this example we have two content facets but hypothesize a three-dimensional space. Levy (1985) pointed out that this is not unusual, because for a circular base it is sufficient to have one facet playing either a polarizing or a modulating role (see Figure 6.3).

A three-dimensional representation of the roles of each facet is given in Figure 6.4.

The fact that facet *A* partitioned the space into a present/future region indicates that the respondents expect changes to take place in the socio, economical and political structures in general.

10. Partial Order Scalogram Analysis

The computer program POSAC provides a two-dimensional representation of a $(n \times p)$ data matrix **P**, called a scalogram. The rows, called profiles or structuples are denoted by $p_1, p_2, \ldots, p_n$, corresponding to cases $1, 2, \ldots, n$, respectively. Each variable has an ordered range and its order is uniform in

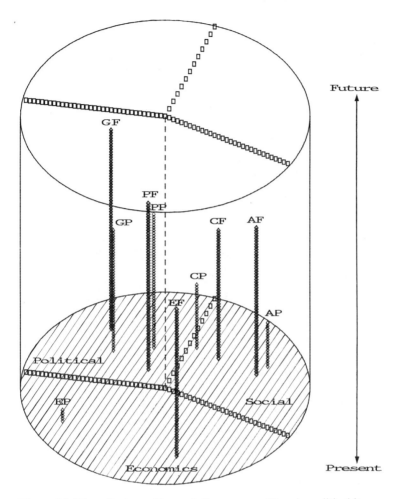

Figure 6.4. The cylindrex of interrelations among 10 socio-political items.

its direction and general meaning for all variables included in the selection (see Section 8).

In a given data set of size n, there will generally be only n^* different profiles, each distinct profile p_k, occuring with frequency f_k, $k = 1, 2, \ldots, n^*$.

The POSAC algorithm is based on the partial order relation that exists between comparable profile pairs. Two profiles p_i and p_k are comparable if p_i is greater than p_k ($p_i > p_k$) or if p_i is smaller than p_k ($p_i < p_k$) or if p_i is equal to p_k ($p_i = p_k$). A profile p_i is greater than p_k if all the elements of p_i have values that are equal to or greater than the values of the corresponding elements of p_k and there exists at least one element in p_i whose value is greater than the value of the corresponding element in p_k. The opposite definition holds in the case of $p_i < p_k$. If the value of an element of p_i is greater than the value of the corresponding element in p_k and at the same time there exists at least one other element in p_i whose value is smaller than the value of the corresponding element in p_k, the profiles are incomparable. Consider for example the profiles $p_1 = (1\ 1\ 2\ 2)$, $p_2 = (1\ 1\ 2\ 1)$ and $p_3 = (2\ 1\ 1\ 1)$. Here $p_1 > p_2$ since each element of p_1 is greater than or equal to the corresponding element of p_2. Profiles p_3 and p_2 are incomparable since the first element of p_3 is greater than, but the third element smaller than the corresponding element of p_2. Comparable profiles satisfy the conditions of reflexivity, antisymmetry and transitivity.

Given a set $\{p_i\}$ of profiles, POSAC seeks a mapping $p_i \to (x_i, y_i)$ where x and y, respectively, denote the horizontal and vertical axes in a two-dimensional space. The mapping of p_i on to a two-dimensional space is subject to:

(1) $p_i > p_k \Leftrightarrow x_i > x_k$ and $y_i > y_k$
(2) p_i incomparable $p_k \Leftrightarrow x_i > x_k$ and $y_i < y_k$
$$\text{or} \quad x_i < x_k \text{ and } y_i > y_k.$$

The above two conditions should hold for as many profile pairs p_i and p_k, i, $k = 1, 2, \ldots, n^*$, as possible.

A loss function incorporating these conditions as well as the frequencies of occurence of each of the n^* distinct profiles is given by Shye and Amar (1985). Optimization of the loss function yields the co-ordinates which are rescaled so that $0 \leq x_i \leq 100$, $0 \leq y_i \leq 100$; $i = 1, 2, \ldots, n^*$. Initial estimates for the vertical and horizontal co-ordinates may be obtained as follows:

Let p_{ij} denote the jth element of p_i, then the score $S(i)$ of profile i in the vertical direction is defined by

$$S(i) = \sum_{j=1}^{p} p_{ij}$$

and the score in the horizontal direction by

$$T(i) = \frac{\sum_{j=1}^{p} [2j - (p + 1)] p_{ij}}{S(i)} + p.$$

For example, if $p_1 = (1\ 2\ 2\ 1)$, then $S(1) = 6$ and

$$T(1) = \frac{(2 - 5)1 + (4 - 5)2 + (6 - 5)2 + (8 - 5)1}{6} + 4 = 4.$$

Suppose $p = 4$ and that each variable (profile element) is an achievement score which varies from 1 to 3. In this case the vertical scores indicate total achievement, while the horizontal scores measure achievement within profiles (e.g., the profiles $(3\ 1\ 1\ 1)$ and $(1\ 1\ 1\ 3)$ have the same S value but T values of 3 and 5, respectively).

The POSAC program also computes Guttman's coefficient of (weak) monotonicity (see Section 9) between the items (variables).

Let $J = X + Y$, where X and Y denote the horizontal and vertical directions (factors) respectively and $L = X - Y$. Together with each item diagram as well as with each trait diagram, weak monotonicity coefficients relating items and traits to the X, Y, J and L directions in two-dimensional space are also printed. This information is very useful for regioning the two-dimensional space. These regions are called regions of contiguity.

Partial order scalogram analysis is now illustrated by means of the following application.

Preferences of 24 Groups for Seven School Subjects

To evaluate the validity of an ability test, school children from the same educational level ($n = 1499$) had to indicate on a 10-point scale how much they like each of the following subjects: first (home) language, a second language, history, geography, mathematics, biology and science.

Table 6.5 gives the data set derived from the measurements on the above seven items. Each profile is described by 4 symbols. The first symbol indicates sex, namely

M—male
F —female.

The second and third symbols denote respectively language group and area where living, namely

AU—Language Group A, urban area
AR—Language Group A, rural area
BU—Language Group B, urban area
BR —Language Group B, rural area.

As fourth symbol the numbers 1, 2, and 3 were used to denote three different socio-economic classes, namely low, middle, and high.

Profiles were obtained by firstly computing the arithmetic mean for each item for each of the 24 subgroups. Secondly, the 25th, 50th and 75th percentiles (q_1, q_2 and q_3) were calculated for the 24 mean preference scores of each school subject. For any given group, values of 1 to 4 were assigned to each item according to the following scheme:

if group mean $\leq q_1$ then assign a value of 1,
if $q_1 <$ group mean $\leq q_2$ assign a value of 2,
if $q_2 <$ group mean $\leq q_3$ assign a value of 3 and
if group mean $> q_3$ assign a value of 4.

The reason why profiles were formed in the above manner, was to facilitate comparison between the 24 groups. A score of 2 or less indicates below average preference and 3 or 4 above average preference ratings.

In Table 6.5 the order of the variables is: profile name (group), sex, language, area where living, socio-economic class, first (home) language, second language, history, geography, mathematics, biology and science.

Table 6.5. Profiles of
preferences of 24
groups for seven school
subjects

```
MAU1 1 1 1 1 4 4 3 4 3 2 3
MAU2 1 1 1 2 2 2 2 2 3 2 2
MAU3 1 1 1 3 1 1 1 3 4 4 4
MAR1 1 1 2 1 4 2 3 3 3 2 3
MAR2 1 1 2 2 2 1 3 1 1 1 3
MAR3 1 1 2 3 1 1 1 1 3 4 4
MEU1 1 2 1 1 3 3 1 4 2 4 2
MEU2 1 2 1 2 2 2 2 3 4 2 2
MEU3 1 2 1 3 2 2 1 2 4 2 4
MER1 1 2 2 1 1 2 3 2 3 1 1
MER2 1 2 2 2 4 1 4 4 4 4 4
MER3 1 2 2 3 1 1 4 3 3 4 4
FAR1 2 1 1 1 4 4 2 2 4 3 2
FAR2 2 1 1 2 3 4 1 2 2 3 2
FAR3 2 1 1 3 2 2 4 4 2 3 1
FAR1 2 1 2 1 4 3 2 2 1 1 2
FAR2 2 1 2 2 3 3 1 1 1 1 1
FAR3 2 1 2 3 1 3 4 4 2 4 3
FEU1 2 2 1 1 4 4 3 3 2 3 3
FEU2 2 2 1 2 3 3 3 1 2 2 1
FEU3 2 2 1 3 3 1 2 1 1 3 3
FER1 2 2 2 1 2 4 2 1 1 1 1
FER2 2 2 2 2 3 4 4 3 4 3 4
FER3 2 2 2 3 1 3 4 4 1 1 1
```

The computer program is as follows.

Input: HUDAP procedure POSAC for obtaining a two-dimensional scaling of the data given in Table 6.5

```
$RUN NAME =   'Preference Scores of 24 Groups' ;
$ITEM NAMES ARE ID SEX LANGGRP AREA SOC_ECON LANG1
              LANG2 HIST GEOGR MATHS BIOL SCIENCE  ;
$DATA         FILE = 'MYFILE1' ;
              FORMAT= '(A4,11F2.0)' ;
$POSAC        NAMES ARE LANG1 TO SCIENCE ;
              LABEL = ID ;
              EXTMAPS =
                    SEX(1) &
                 LANGGRP(1) &
                   AREA(1) &
               SOC_ECON(2) ;

  { External maps are produced for each external trait w.r.t. the
    category specified . The presence of a relevant category
    is indicated by an 'I' symbol and its absence by an 'O' symbol }
$END
```

External maps for sex, language, area where living and socio-economic class are produced by specifying these external traits in the EXTMAPS paragraph.
    Part of the computer output is given below.

Output:

(i) *Details of the different profiles*

```
THERE ARE   26 DIFFERENT PROFILES
```

ID	PROFILE							SCO	FREQ	USER ID
	L	L	H	G	M	B	S			
	A	A	I	E	A	I	C			
	N	N	S	O	T	O	I			
	G	G	T	G	H	L	E			
	1	2		R	S		N			
							C			
							E			
4	4	4	3	4	3	2	3	23	1	MAU1
20	2	2	2	2	3	2	2	15	1	MAU2
12	1	1	1	3	4	4	4	18	1	MAU3
9	4	2	3	3	3	2	3	20	1	MAR1
24	2	1	3	1	1	1	3	12	1	MAR2
18	1	1	1	1	3	4	4	15	1	MAR3
10	3	3	1	4	2	4	2	19	1	MEU1
14	2	2	2	3	4	2	2	17	1	MEU2
15	2	2	1	2	4	2	4	17	1	MEU3
22	1	2	3	2	3	1	1	13	1	MER1
2	4	1	4	4	4	4	4	25	1	MER2
8	1	1	4	3	3	4	4	20	1	MER3
6	4	4	2	2	4	3	2	21	1	FAR1
13	3	4	1	2	2	3	2	17	1	FAR2
11	2	2	4	4	2	3	1	18	1	FAR3
16	4	3	2	2	1	1	1	15	1	FAR1
25	3	3	1	1	1	1	1	11	1	FAR2
7	1	3	4	4	2	4	3	21	1	FAR3
5	4	4	3	3	2	3	3	22	1	FEU1
17	3	3	3	1	2	2	1	15	1	FEU2
21	3	1	2	1	1	3	3	14	1	FEU3
23	2	4	2	1	1	1	1	12	1	FER1
3	3	4	4	3	4	3	4	25	1	FER2
19	1	3	4	4	1	1	1	15	1	FER3
26	1	1	1	1	1	1	1	7	1	NOBODY
1	4	4	4	4	4	4	4	28	1	NOBODY

There are 26 different profiles since the two extreme profiles (1111111) and (4444444) were added by the program.

(ii) *Coefficient of weak monotonicity between the items as well as proportion of profile-pairs correctly represented*

```
            COEFFICIENTS OF WEAK MONOTONICITY
                   BETWEEN THE ITEMS


                          1     2     3     4     5     6     7
                     +- - - - - - - - - - - - - - - - - - - - - - -
                     I
        LANG1      1 I  1.00
                     I
        LANG2      2 I  0.71  1.00
                     I
        HIST       3 I  0.16  0.21  1.00
                     I
        GEOGR      4 I  0.25  0.34  0.76  1.00
                     I
        MATHS      5 I  0.21 -0.02  0.20  0.56  1.00
                     I
        BIOL       6 I  0.11 -0.11  0.24  0.67  0.70  1.00
                     I
        SCIENCE    7 I  0.20 -0.33  0.23  0.41  0.80  0.85  1.00



        PROPORTION OF PROFILE-PAIRS CORRECTLY REPRESENTED
        CORREP COEFFICIENT  .........  0.8031
```

(iii) *Two-dimensional configuration of the scalogram* (Profiles are denoted by (*abcdefg*) where *a* = First language, *b* = Second languge, etc...)

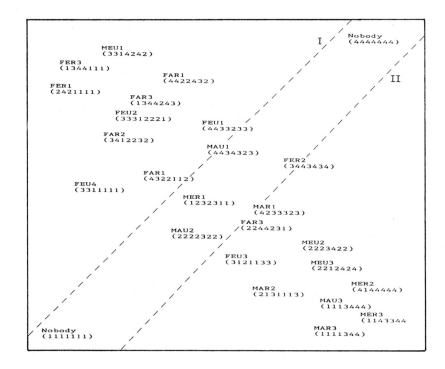

The above representation shows profiles with a higher preference rating for languages than for the science subjects above the dotted line (I) and vice versa for the profiles below the dotted line (II). Profiles between these lines are more or less balanced with respect to the preference scores for the seven school subjects. Overall higher preferences scores are nearer to the top-right profile (4444444) and overall lower preference scores are nearer to the bottom-left profile (1111111).

(iv) *Coefficients of weak monotonicity between items and various directions*

```
COEFFICIENT OF WEAK MONOTONICITY BETWEEN EACH OBSERVED ITEM AND THE FACTORS :
J (I.E. X+Y) , L (I.E. X-Y) , X , Y , P (I.E. MIN(X,Y)) , Q (I.E. MAX(X,Y))

ITEM NAME           J       L       X       Y       P       Q
---------           -       -       -       -       -       -
LANG1      1      0.76   -0.29    0.10    0.56    0.78    0.22
LANG2      2      0.77   -0.90   -0.53    0.97    0.69    0.35
HIST       3      0.83    0.05    0.41    0.34    0.68    0.45
GEOGR      4      0.92   -0.03    0.42    0.46    0.62    0.70
MATHS      5      0.89    0.69    0.88   -0.19    0.65    0.62
BIOL       6      0.83    0.46    0.71   -0.02    0.26    0.81
SCIENCE    7      0.81    0.86    0.94   -0.45    0.39    0.69
```

The above coefficients are useful in that they provide guidelines for regioning the item space. As an example, the coefficient of weak monotonicity (μ_2) between the second language and $L = X - Y$ is -0.90. The regioning of the item space according to this information is shown in the next part of the computer output. Since $\mu_2 = 0.97$ in the Y-direction, the regioning in this direction has even more discriminating power.

(v) *Item diagram for second language*

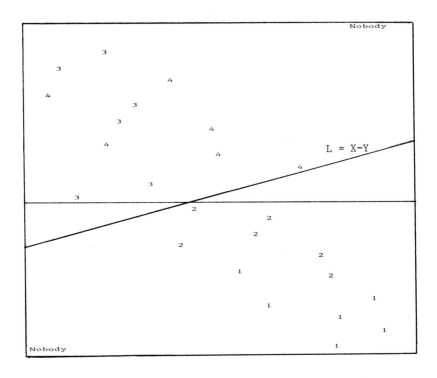

(vi) *Monotonicity coefficients for the external traits*

(a) Sex

```
MU(TRAIT,X)= 0.853                    MU(TRAIT,J)= 0.038
MU(TRAIT,Y)=-0.825                    MU(TRAIT,L)= 0.857
```

(b) Home language

```
MU(TRAIT,X)=-0.082                    MU(TRAIT,J)=-0.483
MU(TRAIT,Y)=-0.184                    MU(TRAIT,L)= 0.053
```

(c) Area where living

```
MU(TRAIT,X)=-0.020                    MU(TRAIT,J)= 0.301
MU(TRAIT,Y)= 0.184                    MU(TRAIT,L)=-0.108
```

(d) Socio-economic class

```
MU(TRAIT,X)= 0.046          MU(TRAIT,J)=-0.307
MU(TRAIT,Y)=-0.240          MU(TRAIT,L)= 0.149
```

From the values of these coefficients it can be deduced that the space for the external trait, sex, may best be regioned in the $L = X - Y$ or X direction. Low values for the coefficients with respect to the other external traits show that no clear regioning exists for these traits. Only the external trait diagram for sex, together with a possible regioning, is therefore given next.

(vii) *External trait for sex*

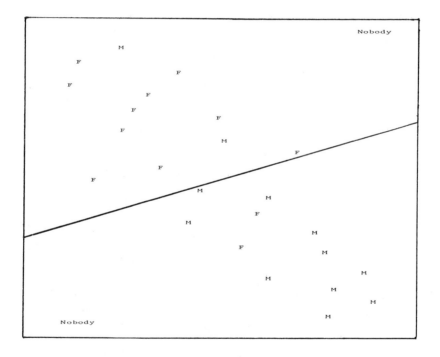

A comparison of this external trait diagram with the two-dimensional configuration of the scalogram shows that the male subgroups are mainly associated with a larger preference for science subjects than for the languages. The regioning into male/female subgroups corresponds to the regioning of the second language item diagram. This implies that male/female differences in school subject preferences are largely explained by their second language preference.

It is interesting to note that preferences for certain subjects are not necessarily associated with achievements in those subjects. A POSAC analysis performed on the actual achievements in the seven school subjects yielded low monotonicity coefficients between the external traits sex, home language, area where living and the X, Y, L and J directions. A high μ_2 value between socio-economic class and the $J = X + Y$ direction indicated the partitioning shown in the external trait diagram for socio-economic class relative to achievements in school subjects given below.

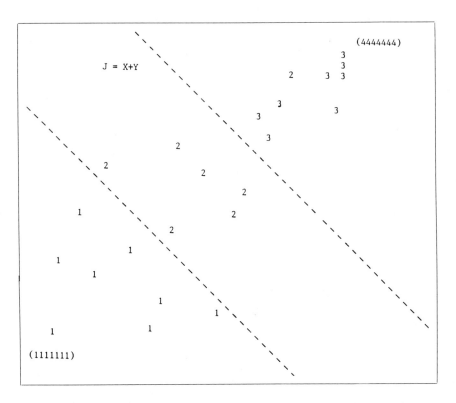

It is seen that Socio-Economic Class 3 is associated with the highest achievements in school subjects and vice versa for Socio-Economic Class 1. The same pattern is, however, not evident in the case of preferences for the various school subjects.

11. General

Apart from the uses of the biplot already discussed, it can also be a useful aid in model identification, in for example, the analysis of variance as described by Gabriel (1981).

For further theoretical information on correspondence analysis and for additional practical applications of this technique, the reader should consult Greenacre (1984).

Sometimes data is obtained in the form of similarities between two sets of objects and furthermore within-set dissimilarities are not available. The multidimensional scaling of this type of data is known as unfolding. The SAS PROC ALSCAL program may be employed for this purpose.

A complete exposition of the theoretical aspects of INDSCAL as well as practical applications is given by Schiffman, Reynolds and Young (1981).

It is possible to perform an analysis using the facet theory approach by making use of any suitable program for non-metric multidimensional scaling. In this case, the researcher will usually not have the facility for specifying facet diagrams, but the elements of the facets can be manually labelled on the graphical representation of the variables.

The interested reader is referred to Canter (1985) for a rigorous treatment of the facet theory approach and of partial order scalogram analysis.

CHAPTER 7

Graphical Representations in Regression Analysis

1. Introduction

In a regression analysis the following graphical techniques are particularly important:

  (i) the scatterplot
 (ii) plots of different types of residuals
(iii) the representation of Mallows' C_k-statistic
 (iv) the construction of confidence and forecast bands
  (v) the ridge trace.

In this chapter these representations will be discussed in the above sequence together with illustrative examples.

In regression analysis it is usually assumed that the dependent variable Y is a random variable of which n values have been observed and that n observations of the p explanatory (independent) variables $X_j, j = 1, 2, \ldots, p$ are also available where $X_{ij}, i = 1, 2, \ldots, n$ indicates the n observations of the jth variable.

In the case of $p = 1$ the so-called simple linear regression model is defined as

$$Y_i = \alpha + \beta X_i + \varepsilon_i.$$

For $p > 1$ the general linear regression model is written as

$$Y_i = \alpha + \beta_1 X_{i1} + \beta_2 X_{i2} + \cdots + \beta_p X_{ip} + \varepsilon_i, \qquad i = 1, 2, \ldots, n.$$

In these models α and the β_j's represent the unknown parameters of the model. Unless otherwise stated the estimates of these parameters are obtained through the method of least squares described, for example, in Draper and Smith (1981) and are represented by $\hat{\alpha}$ and $\hat{\beta}_j, j = 1, 2, \ldots, p$.

The ε_i's are stochastic variables that are not observable and they indicate that the Y_i values are stochastic and cannot be written as an exact linear combination of the explanatory variables. Therefore, the ε_i's are usually known as the error terms.

The estimated Y_i values, represented by $\hat{Y}_i$, are obtained from the regression model through

$$\hat{Y}_i = \hat{\alpha} + \hat{\beta}_1 X_{i1} + \hat{\beta}_2 X_{i2} + \cdots + \hat{\beta}_p X_{ip}, \qquad i = 1, 2, \ldots, n.$$

The multiple correlation coefficient R is defined as the Pearson correlation between Y and $\hat{Y}$. The most familiar measure for evaluating the fit of a general linear regression model is R^2 where $100R^2$ measures the percentage variation of Y explained by the linear regression model.

In a regression analysis the following assumptions are usually made:

(i) The stochastic variables $Y_1, Y_2, \ldots, Y_n$ are mutually independent and each one is normally distributed with constant variance, and
(ii) with means given by

$$\alpha + \beta_1 X_{i1} + \beta_2 X_{i2} + \cdots + \beta_p X_{ip}, \qquad i = 1, 2, \ldots, n.$$

In the case of $p = 1$ it is therefore assumed that the Y values are mutually independent and that each one is normally distributed as $n(\alpha + \beta X_i; \sigma^2)$, $i = 1, 2, \ldots, n$. These assumptions are usually written as follows with regard to the error terms ε_i:

(i) The error terms are mutually independent and each one is normally distributed with constant variance σ^2, and
(ii) each one has a zero mean.

In order to illustrate the graphical techniques which are of importance in regression analysis the data set which is given in Table 7.1 is used repeatedly. This data set represents a random sample of 30 observations with regard to 1978-values of dwellings in Johannesburg obtained from the Central Statistical Service of the Republic of South Africa. The variables Y and X_1 to X_5 respectively signify:

Y = Loan amount granted (in R1000)
X_1 = Valuation of improvements (in R1000)
X_2 = Valuation of site (in R1000)
X_3 = Under-roof area of main building (in sq. metre)
X_4 = Under-roof area of outbuildings (in sq. metre)
X_5 = Age of dwelling where:
     1 = Built before 1940
     2 = Built between 1940 and 1949
     3 = Built between 1950 and 1959
     4 = Built between 1960 and 1969
     5 = Built after 1970

Table 7.1. Observations for Y and X_1 to X_5 for a random sample of 30 dwellings in Johannesburg

Y	X_1	X_2	X_3	X_4	X_5
34.50	32.00	12	214	173	5
15.00	16.00	6	120	40	2
29.00	27.00	10	156	36	1
30.00	29.00	10	280	42	4
29.00	24.50	12	183	62	1
33.00	30.00	11	136	47	3
25.60	22.00	10	228	0	4
15.00	17.00	8	170	42	2
17.50	13.50	6	180	0	1
33.50	28.00	14	159	46	2
60.00	60.00	20	448	62	3
40.00	38.00	17	402	0	4
40.00	73.50	20	523	0	3
15.00	14.00	6	111	107	3
27.00	34.00	8	198	124	5
35.99	23.99	12	194	25	5
20.50	18.50	6	115	34	5
30.00	57.00	25	406	71	3
13.80	12.50	5	133	37	1
21.00	16.50	10	134	0	5
25.00	19.50	12	127	36	2
21.50	15.00	12	98	14	1
32.00	22.00	10	151	45	5
8.90	16.00	8	142	54	4
30.00	31.00	12	237	36	5
22.40	17.90	7	150	102	2
14.90	12.00	5	140	35	1
13.80	13.50	5	112	0	4
33.50	24.50	8	172	51	5
10.90	10.50	4	77	20	5

2. The Scatterplot

Suppose (X_i, Y_i), $i = 1, 2, \ldots, n$ indicate the paired measurements of two variables X and Y. A two-dimensional representation of these pairs of observations is known as a scatterplot. Such plots are particularly useful tools in an exploratory analysis conveying information about the association between X and Y, the dependence of Y on X where Y is a response variable, the clustering of the points, the presence of outliers, etc.

Scatterplots of Y against X_1 to X_5 for the data given in Table 7.1 were obtained by means of the following SAS PROC PLOT program.

Input: SAS PROC PLOT program for obtaining scatterplots

```
DATA    HOUSES ;
INPUT   Y X1-X5 ;
CARDS ;
 34.50 32.00 12 214 173 5
 15.00 16.00  6 120  40 2
 29.00 27.00 10 156  36 1
               .
               .
               .
 13.80 13.50  5 112   0 4
 33.50 24.50  8 172  51 5
 10.90 10.50  4  77  20 5
 ;
PROC    PLOT ;
PLOT    Y*X1 Y*(X2-X5) = '*' / VPOS=24 HPOS=40 ;
TITLE Scatterplots of Y against X1 to X5 ;
```

In the above program a plotting symbol '*' was specified for the scatterplots of Y against X_2 to X_5. The size of the figures can be determined by the VPOS and HPOS options. In this program they were chosen so as to produce fairly square plots. Only the scatterplots of Y against X_1 and of Y against X_2 are given below.

Output:

(i) *Scatterplot of Y against X_1*

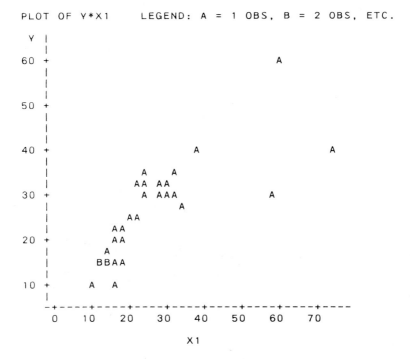

(ii) *Scatterplot of Y against X_2*

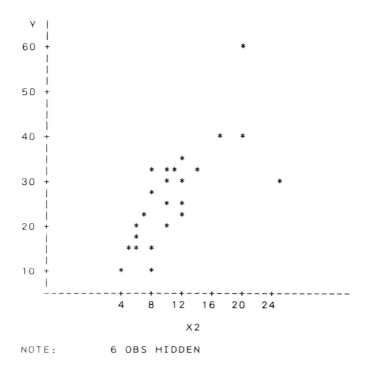

NOTE: 6 OBS HIDDEN

Both these scatterplots give an indication of a positive association between the Y and X variables. Comparing the two plots with regard to clustering of points it can be seen that the plotting symbols used in the plot of Y against X_1 give an indication of any overlapping points. In contrast the plot of Y against X_2 with plotting symbol '*' does not allow for an identification of overlapping points. In fact in this latter plot, six observations are hidden.

Another approach to solving the overlapping problem is presented by Cleveland and McGill (1984b). These authors divide the plotting region (which is chosen to be square) into a large number of subregions or cells of equal size. The number of points in each cell are then obtained and portrayed by symbols called sunflowers. A single dot, for example, represents a count of 1, a dot with two line segments a count of 2, a dot with three line segments represents a count of 3 etc.

The above authors present a number of ingenious methods by which visual discrimination between points can be enhanced. They conclude that in general, simple shapes such as circles or squares provide less discrimination than do different methods of filling these shapes. Based upon some empirical observations they also suggest that in making a scatterplot, one should control the point cloud size and not allow the cloud of points to be either too far or too close to the frame of the plot. In the case of possible outliers they recommend making two scatterplots, with and without outliers since outliers can lead to reductions in the point cloud size by forcing the vast majority of points into a small region.

When making a scatterplot Tukey (1977) considers the following aspects to be important:

(a) A scatterplot of Y against X on tracing paper placed on top of graph paper often provides better insight into the form of the regression equation that should be used. A representation on tracing paper has the advantage that any distraction caused by the underlying grid of the graph paper is eliminated.
(b) In a scatterplot not too many scale values should appear on the final representation since the plotted points should show up as freely and as clearly as possible. The horizontal scale values can be placed at the top of the figure so that when tracing the scatterplot the hand does not obscure the scale values.
(c) The plotted points in a scatterplot should never be joined point for point. The whole idea of a scatterplot is to be able to observe the distribution of the points without the distracting effect of a zig-zag representation.

A. The Scatterplot in Simple Linear Regression ($p = 1$)

In this case a linear regression analysis is carried out on the pairs of observations (X_i, Y_i), $i = 1, 2, \ldots, n$.

In a meaningful regression analysis the explanatory variable (X) should be highly correlated with the dependent variable (Y). The correlation coefficient r only measures the linear relation between X and Y and is not a reliable measure of association if the relation between X and Y is nonlinear. A scatterplot of Y against X will usually give a good indication of the degree of linearity between X and Y as well as of alternative regression models that may be considered in the case of a nonlinear relationship between X and Y.

The figures below indicate a number of scatterplots. In Figure (a) there is a strong linear relationship between X and Y so that $Y = \alpha + \beta X + \varepsilon$ is a meaningful model for the data.

In Figure (b) there is little or no relation between X and Y. In this case the correlation coefficient r will be approximately zero. A regression analysis would make little sense in this case since X has no explanatory power in respect of Y.

Although the correlation coefficient r in Figure (c) would be very small, there is nevertheless a strong nonlinear relationship between X and Y. This scatterplot suggests $Y = \alpha + \beta X + \gamma X^2 + \varepsilon$ as the regression model.

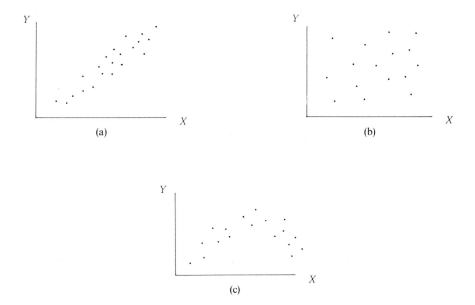

Any regression analysis should be preceded by a scatterplot. It is not sufficient merely to study the R^2-value and then to evaluate the model on this basis. Anscombe (1973) provides four different data sets of the same size, all of which with a regression model of the form $Y = \alpha + \beta X + \varepsilon$ produce the same $\hat{\alpha}$, $\hat{\beta}$ and R^2-values. From the scatterplots of the four data sets in Figure 7.1 it is evident however, that fitting a straight line is not meaningful in all four cases.

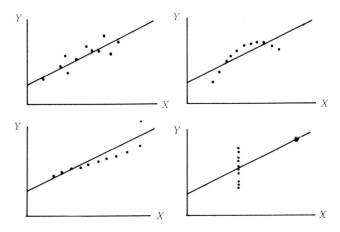

Figure 7.1. Scatterplots of Anscombe's (1973) four data sets.

Table 7.2. Nonlinear functions that can be written in a linear form by means of a transformation

Type	Nonlinear form	Transformation	Linear form	Presentation
Hyperbolic	$Y = X/(\alpha X - \beta)$	$Y' = 1/Y, X' = 1/X$	$Y' = \alpha - \beta X'$	Figure 7.2a
Exponential	$Y = \alpha \exp(\beta X)$	$Y' = \log Y$	$Y' = \log \alpha + \beta X$	Figure 7.2b
Power function	$Y = \alpha X^{\beta}$	$Y' = \log Y, X' = \log X$	$Y' = \log \alpha + \beta X'$	Figure 7.2c
Logarithmic	$Y = \alpha + \beta(\log X)$	$X' = \log X$	$Y = \alpha + \beta X'$	Figure 7.2d
Logit	$Y = \dfrac{\exp(\alpha + \beta X)}{1 + \exp(\alpha + \beta X)}$	$Y' = \log(Y/(1 - Y))$	$Y' = \alpha + \beta X$	Figure 7.2e

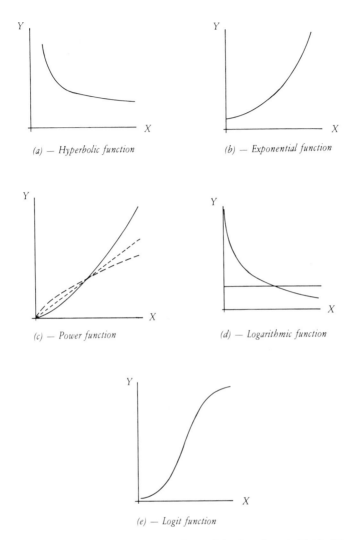

(a) — Hyperbolic function

(b) — Exponential function

(c) — Power function

(d) — Logarithmic function

(e) — Logit function

Figure 7.2. Graphical representations of the functions in Table 7.2.

A regression model is called linear if the model is linear in respect of the parameters. The model $Y = \alpha + \beta X + \varepsilon$, for instance, is linear whereas the model $Y = \alpha X^\beta + \varepsilon$ is nonlinear.

Various nonlinear models can be written in a linear form through appropriate transformations. Table 7.2 gives a number of nonlinear models together with the transformations for converting them into linear models. A reference to the representation in Figure 7.2 of the functional relation is also given in each case. These representations could be of use if the model were to be chosen on the basis of a scatterplot.

As an example of a scatterplot indicating a possible nonlinear regression model which through an appropriate transformation can be written in a linear form consider the plot given below in which the variable, fetal mass (in grams) is depicted against the variable, number of weeks since conception. The data were obtained from patient files from private medical practices in two cities in the Republic of South Africa. The fetal mass was calculated by a formula in which sonar measurements were used.

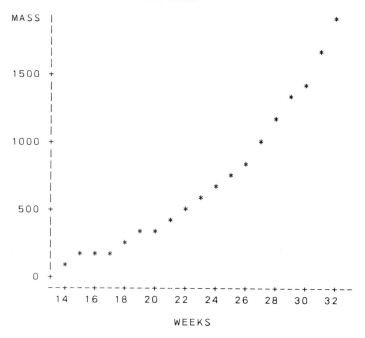

Since the form of the plot in the above figure resembles that of an exponential function [see Figure 7.2(b)] the logarithm of fetal mass was taken (log (FETAL MASS) = Y) and depicted against weeks since conception, yielding the following figure which clearly shows a near linear relationship.

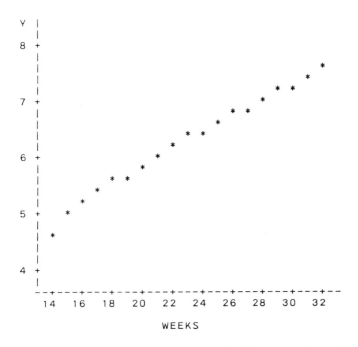

WEEKS

As has been illustrated, scatterplots are a valuable tool in determining the form of the functional relationship between a response variable, *Y* and a predictor *X*. In data sets where the response variable displays a considerable amount of variation this task is often not as simple as it would seem. This difficulty of choice is evident from the following application: Occupational choice theory as developed by Holland (1985) can be used to classify individuals into six occupational groups which makes occupational guidance and placing considerably easier. In the scatterplot given below an index of interest differention (*Y*), based on Healy and Mourton's (1983) index, is plotted against a measure of enterprising behavior (*X*) for a sample of 206 boys from Educational Level III.

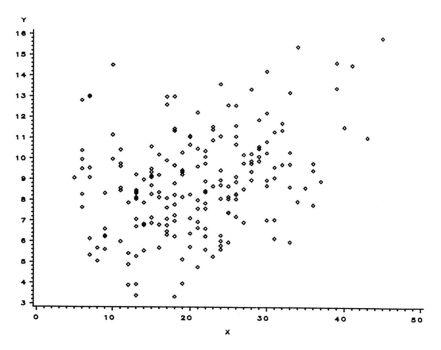

A technique which leads to a more satisfactory model choice in cases such as the one mentioned above is that of scatterplot smoothing.

Smoothed scatterplots are obtained by plotting the points (X_i, Y_i^*) where the Y_i^*-values are the smoothed values. Although a large number of ways exist by which one can obtain smoothed values of Y, only the symmetric nearest neighbor smoothing procedure is discussed in this section since it is computationally simple.

A symmetric nearest neighbor smoothing entails the choosing of a window or a span by which the neighborhood of X_i is specified. A span of between 10 and 50 percent of the observations is usually chosen and 30 percent seems to work reasonably well in most cases. Choosing a span of 30 percent means that the smoothed value Y_i^*, is obtained from the $0.15n$ values of Y_i to the left of X_i and the $0.15n$ values to the right of X_i. In this procedure the neighborhoods are truncated towards the ends of the X_i.

The Y_i^* can be obtained by calculating the average value of the Y_i contained in each successive span or by fitting straight lines to all the (X, Y) values in the neighborhood of X_i. The predicted value $\hat{Y}_i$ at X_i is then taken as the smoothed value Y_i^*. This so-called running lines smoother tends to produce less bias on the boundaries and can be refined further, as discussed in Cleveland and McGill (1984b), by fitting the straight lines by a weighted least squares regression in which points close to X_i receive a larger weight than points further away from X_i.

As an illustration a running lines smoother with a window size of 30 percent was applied to the occupational choice application mentioned earlier.

Input: SAS PROC MATRIX program for scatterplot smoothing

```
DATA    CAREERS ;
INPUT   DIF E @@ ;
CARDS ;
 8.04 26    8.28 15    9.49  6    7.08 13    7.15 15    5.99 11    7.26 18
 9.77 27    9.20 16    8.24 18   11.37 32   14.51 10   11.07 26    8.95 24
 9.05  5    9.75 22    7.12 20    9.09 15   11.29 31   10.27 30   11.08 24
10.09 29   13.60 24    7.39 25    4.88 12    8.71 24    6.72 13    3.89 12
 8.95 30   10.69 26   11.71 32   14.22 30    8.37 16    8.66 25    7.20 26
15.40 34    7.09 17   10.67 20   12.22 21    5.13 19    6.27  9    6.78 17
 8.96 28   10.19 16    8.39 13   13.19 33    5.55 14    7.39 25    9.43 24
 9.44 36   10.30 33   10.29 19    8.26 26    9.35 15    5.42 12   14.52 41
 9.22 19    8.15 25    5.27 23   12.56 26    5.67  8    9.37 28   11.09 20
 6.32 17    7.63  6    8.12 13    9.75 28    5.96 25    8.06 24    6.12  7
 9.61 21    6.81 16    6.15 31    5.73 20    8.26 20   11.54 23    7.58 22
10.08 22    8.57 11   11.05 20    9.71 32    6.48 17    8.45 11    8.98 14
12.60 17    6.99 18    8.43 22    6.79 14    9.95  6    9.43 19    7.84 22
 6.60  9    8.66 23   10.48 29    9.87 29    7.78 36    9.91 17   12.98 17
 8.04 13   10.38  6   12.58 25   12.97  7    4.76 21    8.64 33    6.85 14
 9.22 13   10.58 29    3.93 13   10.42 11    9.50 15    5.33  7    9.87 28
 8.70 15    7.78 18    6.65 21    7.58 21    8.32 13    8.60 35    8.52 27
 9.58 31   10.82 30    5.69 16    8.80 18   11.44 18    8.04 25    9.67 18
 9.16 15    8.14 19    6.24 22    8.87 26    6.02 24    5.60 22    6.22  9
 7.63 17    5.28 13   11.40 23    5.98 33    9.35 19   10.37 22    6.24 22
 7.87 14    7.06 31   12.80  6    8.32  9   10.89 23    7.04 30    5.61 24
 7.38 25    6.92 27   11.14 10    9.78 36    6.62 22    8.92 37   11.03 43
 8.37 22    8.69 30    6.95 21    5.04  8   11.53 40    8.09 17   11.87 29
12.20 30    3.39 13   13.01  7    3.33 18    3.97 19    6.24 18    7.94 34
 9.75 11    9.97 10    7.87 12    6.89 15    5.79 24    8.46 13    9.12 26
10.58 15   12.99 18    8.26  6    9.61 11    9.07  7    6.28 24    6.38 20
 9.73 33    9.54  7    5.61  9    7.99 21    9.07 31    9.67 18   13.35 28
 8.13 16   11.34 18   13.41 39   10.48 21   15.82 45   14.64 39    8.90 22
10.18 27    8.28 26   11.59 26   10.23 28
;
PROC    SORT ;
        BY E ;                          * Sort independant variable ;
PROC    MATRIX ;
        FETCH XX ;
        NOBS = NROW(XX) ;
* SET   X , Y & YHAT ;
        X = XX( ,2) ;
        Y = XX( ,1) ;
        YHAT=J(NOBS,1,0) ;
DO NTRY = 1 TO 2 ;                       * Repeat smoothing procedure twice ;
        HNSPAN = INT(0.15*NOBS)  ; * Window size 30% of data ;
        IF NTRY = 2 THEN HNSPAN = HNSPAN #/3 ; * Window size 10% ;
        IF HNSPAN  = 2.0*INT(HNSPAN#/2) THEN HNSPAN = HNSPAN - 1 ;
        HN1 = HNSPAN - 1 ;
PRINT   HNSPAN ;
        SXIL = 0 ;
        SYIL = 0 ;
        SXIXIL= 0 ;
        SXIYIL = 0 ;
IF NTRY = 2 THEN Y = YHAT ;
* INITIATE SUMS ;
DO K = 1 TO HN1 ;
        XK = X(K,1) ;
        YK = Y(K,1) ;
        SXIL    = SXIL    + XK ;
        SYIL    = SYIL    + YK ;
        SXIXIL  = SXIXIL  + XK*XK ;
        SXIYIL = SXIYIL + XK*YK ;
END ;  * K -LOOP ;
```

```
* OBTAIN MOVING STRAIGHT LINES ;
       L1 =   HNSPAN + 1 ;
       L2 =   NOBS - HNSPAN + 1   ;
       L3 =   L2 + 1  ;
DO L = 1 TO NOBS ;
       LH      = HNSPAN + L - 1 ;
       LL      = L - HNSPAN ;
       IF L <= HNSPAN THEN  NL  = HNSPAN + L - 1 ;
       IF  L >= L1  AND L <=L2 THEN  NL  = 2.0* HNSPAN - 1 ;
       IF L >= L3   THEN  NL = NOBS + HNSPAN -L ;
       XLL = 0 ;
       YLL = 0 ;
       IF LL >= 1 THEN XLL = X(LL,1) ;
       IF LL >= 1 THEN YLL = Y(LL,1) ;
       IF LH <= NOBS THEN XLH = X(LH,1) ;
       IF LH <= NOBS THEN YLH = Y(LH,1) ;
       IF LH >  NOBS THEN XLH = 0 ;
       IF LH >  NOBS THEN YLH = 0 ;
       SXIL    = SXIL    + XLH        -XLL    ;
       SYIL    = SYIL    + YLH        -YLL    ;
       SXIXIL = SXIXIL   + XLH*XLH - XLL*XLL ;
       SXIYIL = SXIYIL   + XLH*YLH - XLL*YLL ;
* CALCULATE Y-HAT ;
       W1 = SXIYIL - SXIL*SYIL#/NL ;
       W2 = SXIXIL  - SXIL*SXIL#/NL ;
       WL = 0 ;
       IF(W2 > 0 )  THEN WL = W1#/W2 ;
       YHAT(L,1) = (SYIL#/NL) + (X(L,1) - (SXIL#/NL))*WL ;
END ;  * L - LOOP ;
END ;  * NTRY-LOOP ;
       SUMM =  XX || YHAT ;
       FREE XX X Y YHAT ;

PRINT SUMM ;
OUTPUT SUMM DATA=SCATTER (RENAME=(COL1=Y COL2=X COL3=P )) ;
TITLE .C=BLUE Scatterplot Smooth;
PROC   GPLOT ;
       PLOT   Y*X=2 ;                        * Plot of raw data only ;
       PLOT   Y*X=2    P*X=3 /OVERLAY ;
GOPTIONS HSIZE=7 VSIZE=7 SPEED=7 ;
       SYMBOL1 V=STAR C=BLACK ;
       SYMBOL2 V=DIAMOND C=RED ;
       SYMBOL3 I=JOIN  C=GREEN ;
FOOTNOTE1 .C=BLACK 'Occupational choice data' ;
FOOTNOTE2 .C=BLUE  'Window size is 30% of the data' ;
```

Output:

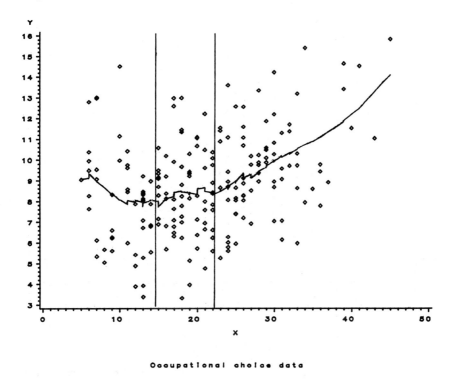

Occupational choice data

In the above output the window size of 30 percent is graphically displayed for $X = 18$. If the smoothed curve depicts some jagged peaks it can be smoothed again by applying the smoothing procedure to the predicted Y-values, $\hat{Y}$ obtained in the first application. In such a second application a smaller window size is recommended. The scatterplot given next shows the effect of such a second application of the smoothing procedure and makes the choice of a probable functional relationship between X and Y considerably easier than was the case for the original unsmoothed scatterplot.

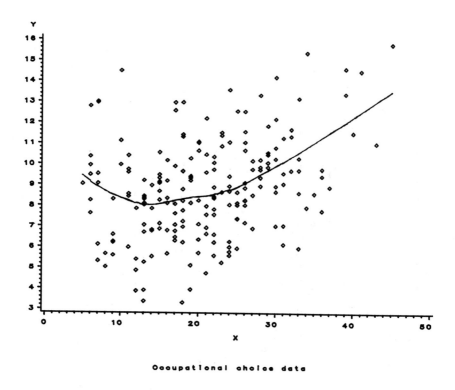

Ooaupatlonal ahoioe data

Scatterplot smoothing is also useful in smoothing a scatterplot arising from binary data. In a 1985-study aimed at determining how export orientated South African manufacturing concerns are a random sample of 194 firms was obtained. Part of the study entailed determining whether a relationship exists between making use of services rendered by the South African Foreign Trade Organisation (SAFTO) and the turn-over of the firm (X) where the binary variable Y is defined as:

$$Y = \begin{cases} 1 & \text{if firm makes use of SAFTO's services} \\ 0 & \text{if not.} \end{cases}$$

Since turnover varied between zero and R500 million, X was transformed to $\ln(X)$. The data set is given below followed by the smoothed scatterplot.

0 12.87	1 15.67	1 13.83	0 13.53	1 14.12	1 15.34	1 14.68	0 14.88
1 16.02	0 11.73	1 14.49	1 12.82	0 13.27	1 12.19	0 15.80	1 15.58
0 13.53	0 13.68	0 10.10	1 14.40	1 13.78	1 13.83	0 11.66	0 13.83
0 14.16	1 13.61	0 14.05	0 12.43	0 13.72	0 9.94	0 12.46	0 10.46
0 11.78	1 13.80	0 11.46	1 15.14	1 14.38	0 12.05	1 17.39	0 13.28
1 13.98	0 14.48	0 14.47	0 12.40	0 13.57	0 13.07	0 10.87	1 13.28
1 13.00	0 12.75	0 13.23	1 12.39	1 16.42	1 16.72	1 12.44	0 10.07

1	15.41	1	9.44	0	11.70	0	14.12	0	11.76	0	12.06	1	15.15	0	13.79
0	12.40	0	11.00	0	14.45	1	16.48	0	10.71	1	15.76	1	14.19	0	12.51
0	12.06	0	14.08	1	16.79	0	14.96	1	15.76	0	12.38	1	15.29	1	11.67
0	12.79	0	11.89	1	14.83	1	12.31	0	14.32	0	12.23	1	13.34	0	11.17
1	15.69	1	15.26	0	13.42	0	15.68	1	15.97	1	13.34	0	12.84	1	15.63
1	16.66	1	14.92	0	9.34	1	16.75	1	14.20	0	10.41	1	13.67	1	14.03
1	12.37	1	17.65	0	14.69	1	16.53	1	14.30	0	15.16	0	13.17	1	16.11
1	14.01	1	14.31	0	11.81	0	14.30	1	15.66	1	13.14	0	9.39	0	11.44
1	16.60	1	12.54	1	15.42	1	16.56	0	13.44	0	12.72	0	12.14	1	12.61
1	15.61	0	13.38	0	12.63	1	12.55	1	14.06	1	15.91	1	15.86	1	15.56
1	17.54	1	16.30	1	15.07	1	14.62	1	15.58	1	16.72	1	16.24	0	15.29
1	15.30	1	14.22	1	15.91	1	15.90	1	15.17	1	16.25	1	18.40	0	14.96
1	14.65	1	16.74	1	14.67	1	19.05	1	13.88	1	16.61	1	18.22	1	15.32
1	17.11	1	15.13	1	15.29	1	15.01	1	14.08	1	16.63	1	17.25	1	14.85
1	16.77	1	16.11	1	18.56	1	16.03	1	16.12	1	14.17	1	14.44	1	15.08
1	14.46	1	16.11	1	16.11	1	17.08	0	13.29	1	16.47	0	11.64	1	17.48
0	15.39	0	15.88	1	11.30	1	11.51	1	15.27	1	14.51	1	15.82	1	14.69
0	12.49	1	16.38												

Scatterplot Smooth

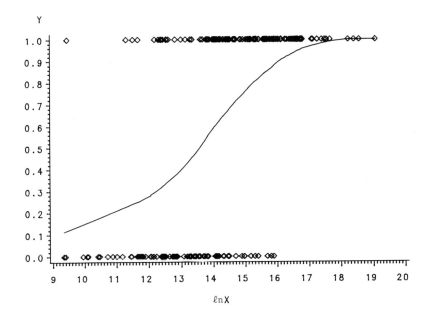

Export data

Window Size is 30% of the data

From this smoothed scatterplot a positive relationship between Y and $\ln(X)$ is evident. This implies that the probability of making use of SAFTO's services increases as turnover increases.

B. The Scatterplot in General Linear Regression ($p \geq 2$)

In this case a scatterplot of Y against each of the p explanatory variables can be made. Two objections can be raised against this approach. If p becomes large the simultaneous interpretation of a large number of plots may lead to confusion. Such scatterplots can also be misleading since the structure present in the original $(p + 1)$-dimensional space is not necessarily reflected by the individual structures present in the scatterplots of pairs of variables. Various techniques in which an attempt is made to represent the original $(p + 1)$-dimensional data in a reduced dimension while retaining the structure of the original space as far as possible, have already been discussed in Chapter 6.

The question whether Y and the X_j, $j = 1, 2, \ldots, p$ are correlated and whether the X_j are mutually uncorrelated, can be investigated inter alia by analyzing the matrix of partial correlation coefficients.

In the multivariate case scatterplots therefore have a limited value with regard to the choice of model, so that this problem is usually approached by means of more general analytical techniques. If a single explanatory variable X_j plays a relatively dominant role compared to all the other explanatory variables, the corresponding (X_j, Y) scatterplot can be of value.

3. Residual Plots

Residual analysis is of vital importance in any regression analysis. A residual analysis entails the careful evaluation of the differences between the observed values and the estimated values of the dependent variable after fitting a regression model to the data.

Residual plots are used inter alia with a view to identifying any undetected tendencies in the data, as well as outliers and fluctuation in the variance of the dependent variable. However, interpretation of such residual plots requires great care on account of the possible degree of subjectivity involved therein.

The following residuals will be discussed in this section:

(a) Least squares or ordinary residuals
(b) Studentized residuals
(c) Partial residuals
(d) Deleted residuals and
(e) Recursive residuals.

With the exception of recursive residuals, the BMDP-1R and 9R multiple linear regression programs are used to calculate residuals and to obtain residual plots for a multiple linear regression with regard to the data set given in Table 7.1. For this data set a multiple linear regression analysis leads to the following prediction equation:

$$\hat{Y} = 6.169 + 0.344X_1 + 0.764X_2 - 0.001X_3 + 0.014X_4 + 0.819X_5.$$

A. Least Squares or Ordinary Residuals

Least squares residuals are defined as the difference between the observed value Y_i and the predicted value $\hat{Y}_i$ and are given by:

$$e_i = Y_i - \hat{Y}_i, \qquad i = 1, 2, \ldots, n.$$

These residuals must not be confused with the errors ε_i which were considered in the introduction to this chapter. By assumption the errors are uncorrelated random variables with zero means and common variance σ^2. The least squares residuals also have zero means but generally they all have different variances and are also correlated.

If it is assumed that the errors are independently distributed as $n(0; \sigma^2)$ then the best estimator of the unknown variance σ^2 is given by:

$$s^2 = \sum_i e_i^2 / (n - p - 1).$$

From the above, ε_i/σ can thus be estimated by e_i/s which represents a standardised residual. Thus the e_i/s can be used to evaluate the assumption that the ε_i/σ are distributed as $n(0; 1)$. Strictly speaking, on the assumption that the ε_i/σ are distributed $n(0; 1)$, e_i/s possesses a t distribution with $(n - p - 1)$ degrees of freedom, but if $(n - p - 1)$ is relatively large the normal approximation of the t distribution can be used to confirm the assumption of normality of the error terms. Since approximately 95% of standard normal deviates are expected to lie within the interval $(-2; 2)$, the assumption of normality of the errors can be evaluated by judging the (e_i/s)-values in respect of this interval.

Draper and Smith (1981, p. 144) point out that although the above method works satisfactorily in practice, theoretically speaking the (e_i/s) values are not mutually independent for small values of $n - p - 1$ and neither do they possess the same variance. A theoretically more refined method is discussed by these authors.

Various least squares residual plots will now be discussed. Consider firstly the assumption that the error terms are independently distributed as $n(0; \sigma^2)$. If the model is specified correctly the residuals e_i should confirm the above assumption.

Suppose $Y_1, Y_2, \ldots, Y_n$ are collected in a specific order (say over time), then a time series representation of the residuals should indicate a random distribution around the zero line, provided the error terms are uncorrelated. In the figure below, the residuals are depicted against the observation order t. From the figure it is evident that the residuals are not randomly distributed about the zero line but reveal a definite tendency over time, which indicates that the error terms are correlated. The residual plot below is an example of so-called positive auto-correlation which frequently occurs in economic time series models.

Such a time series representation can also be used (if applicable) to check whether the variances of the error terms increase or decrease over time (on the assumption of independence) and whether linear or quadratic terms in time should be included in the model.

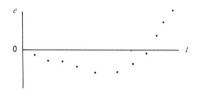

If the observations $Y_1, Y_2, \ldots, Y_n$ represent cross-section data (for example, the data in Table 7.1) the assumption of homoscedasticity (equal variances of the error terms) is usually investigated by plotting e against $\hat{Y}$. It can be shown that the correlation coefficient between e and $\hat{Y}$ is zero for regression equations for which $\alpha \neq 0$. A plot of e against $\hat{Y}$ should therefore yield a random distribution of points about the zero line. (Since e and Y are usually correlated, e should not be represented against Y.)

In the figure below e is represented against $\hat{Y}$. The distribution of residuals about the line $e = 0$, confirms the assumption of homoscedasticity.

In the next two figures the plot of e against $\hat{Y}$ respectively takes on a diverging and converging conical shape.

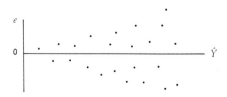

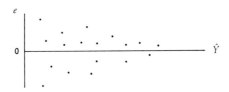

In both cases such a representation indicates that the assumption of the constant variances of the error terms is not valid. A possible solution to this problem is to carry out a weighted least squares analysis or to stabilize the variances prior to estimating the parameters by applying a transformation to the Y values.

The transformations most commonly used are:

(i) $\sqrt{Y}$—the Y's are counts (Poisson scores).
(ii) $\log(Y)$—the Y's vary from very small positive to very large positive.

Weisberg (1980) provides a list of transformations which may be used to stabilize the variances in the case of heteroscedasticity.

In evaluating the assumption of homoscedasticity it may also be useful to represent e^2 against $\hat{Y}$ since e^2 represents the contribution of a given response to the error sum of squares used in estimating the unknown variance $\sigma^2 (s^2 = \sum_i e_i^2/(n - p - 1))$. Such a representation can point out relations between e and $\hat{Y}$ more clearly.

In summary, the assumption of homoscedasticity can therefore be studied by plotting e against $\hat{Y}$ and e^2 against $\hat{Y}$.

Both these plots can be obtained by means of the BMDP-1R regression program by specifying the RESIDUAL statement in the PLOT paragraph. The size of the plots is controlled by means of the SIZE statement. In these representations e was obtained from the full regression model

$$Y = \alpha + \beta_1 X_1 + \beta_2 X_2 + \beta_3 X_3 + \beta_4 X_4 + \beta_5 X_5 + \varepsilon$$

applied to the data given in Table 7.1.

Input: BMDP-1R-program for plotting least squares residuals

```
/PROBLEM      TITLE IS 'REGRESSION OF Y ON X1-X5 '.
/INPUT        VARIABLES ARE 6.
              FORMAT = '(2F6.2,F3.0,2F4.0,F2.0)'.
/VARIABLE     NAMES ARE Y,X1,X2,X3,X4,X5.
/REGRESS      DEPENDENT=Y.
              INDEPENDENT=X1,X2,X3,X4,X5.
/PLOT         RESIDUAL.
              NORMAL.
              VARIABLE=X1,X2,X3,X4,X5.
              SIZE=30,20.
/END
 34.50 32.00 12 214 173 5
 15.00 16.00  6 120  40 2
 29.00 27.00 10 156  36 1
           .
           .
           .
 13.80 13.50  5 112   0 4
 33.50 24.50  8 172  51 5
 10.90 10.50  4  77  20 5
/*
//
```

Output:

(i) *Plot of e against $\hat{Y}$*

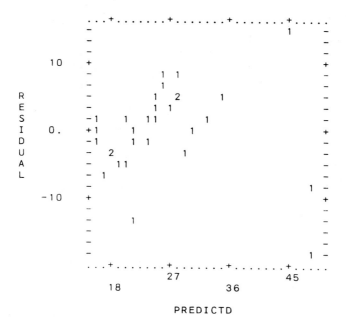

(ii) *Plot of e^2 against $\hat{Y}$*

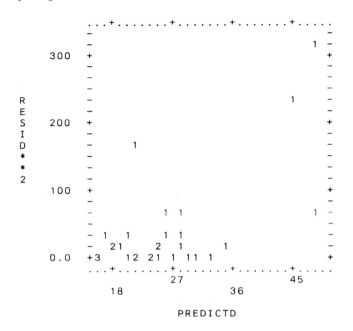

In both plots it appears that the assumption of homoscedasticity may not hold since there are indications that larger values of e tend to accompany larger values of $\hat{Y}$. If considered serious enough, this type of problem could be solved by carrying out a weighted least squares analysis or by transforming Y through an appropriate transformation, possibly $\log(Y)$.

Least squares residuals can also be used to investigate the assumption of normality of the error terms. A regression analysis usually involves the testing of hypotheses and the construction of confidence intervals using techniques based on the assumption of normality. The validity of this assumption is usually investigated by testing whether the residuals can be regarded as n observations from a normal population with a zero mean and variance σ^2. The methods described in Chapter 3 can be used here.

The required normal probability plot can be obtained by specifying the NORMAL statement in the PLOT paragraph of the BMDP-1R program.

Output: Illustration of the NORMAL statement in the BMDP-1R program

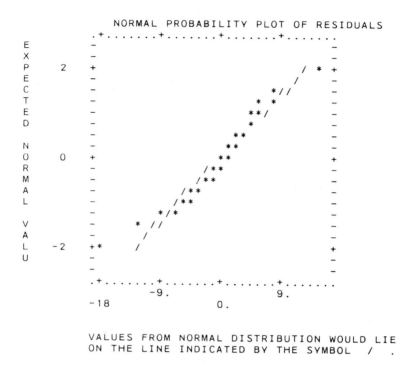

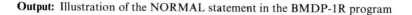

The above normal probability plot indicates a heavy tailed distribution (see Chapter 3) so that the distribution of the errors may deviate slightly from normality.

Consider secondly, the use of least squares residual plots in examining the correctness and relevance of the regression equation.

Specifying the correct regression equation involves, amongst other things, two important matters. Firstly, all the relevant variables should be included in the regression equation and secondly, the correct functional form of each explanatory variable (or predictor) should be used in the prediction equation.

The question of whether X_j is specified correctly in the prediction equation can be investigated by plotting the ordinary residuals against X_j. If the assumptions in respect of the errors are valid and a linear model is specified in $X_1, X_2, \ldots, X_p$, it can be shown that the correlation between e and each predictor X_j is zero. In a plot of e against X_j the residuals should therefore be distributed randomly about the $e = 0$ line. Such a random distribution of residuals would then indicate a correct specification of X_j in the prediction equation.

The representation of e against X_j in the figure below, for instance, indicates that further terms in X_j (e.g., a quadratic term) should be included in the prediction equation. Great care must however be taken when including such a quadratic term due to the subsequent problem of possible multicollinearity. This aspect is discussed further in Section 6.

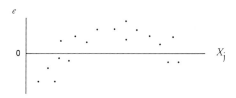

For the data set in Table 7.1 the plots of e against X_1, X_2, X_3, X_4 and X_5 can be obtained by specifying the VARIABLE $= X_1$, X_2, X_3, X_4, X_5 statement in the PLOT paragraph of the BMDP-1R program. In order to illustrate this, only the plot of e against X_2 is given here.

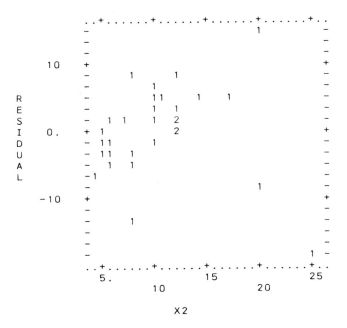

Although the distribution of the residuals around the $e = 0$ line in the above figure does not appear to be entirely random, specifying X_2 in linear form in the regression equation is seemingly correct.

Residual plots can also be used to decide whether an interaction term X_iX_j, should be included in the model. On the usual assumptions of a regression analysis the effect of any predictor X_i on the response variable Y should, ideally speaking, not be influenced by changes in another predictor X_j. (This follows since the predictors should not be correlated.) A plot of the residuals which has been obtained from a prediction equation in which X_iX_j does not occur, as against $X_i X_j$, should, as before, produce randomly distributed residuals about the line $e = 0$. Another pattern in this representation would mean that the residuals and the interaction terms are dependent and that $X_i X_j$, or a function thereof should be included in the model.

Such a plot can be obtained with the BMDP-1R program and is illustrated on the basis of the interaction term $(X_1 X_5)$. This represents a plausible interaction between valuation of improvements (X_1) and age of dwelling (X_5).

Input: BMDP-1R-program for the plot of e against $X_1 X_5$

```
/PROBLEM      TITLE IS 'REGRESSION OF Y ON X1-X5 '.
/INPUT        VARIABLES ARE 6.
              FORMAT = '(2F6.2,F3.0,2F4.0,F2.0)'.
/VARIABLE     NAMES ARE Y,X1,X2,X3,X4,X5,X6.
              ADD=1.
              AFTER.
/TRANSFORM    X6=X1*X5.
/REGRESS      DEPENDENT=Y.
              INDEPENDENT=X1,X2,X3,X4,X5.
/PLOT         VARIABLE=X6.
              SIZE=30,20.
/END
 34.50 32.00 12 214 173 5
 15.00 16.00  6 120  40 2
 29.00 27.00 10 156  36 1
                .
                .
                .
 13.80 13.50  5 112   0 4
 33.50 24.50  8 172  51 5
 10.90 10.50  4  77  20 5
/*
//
```

Output:

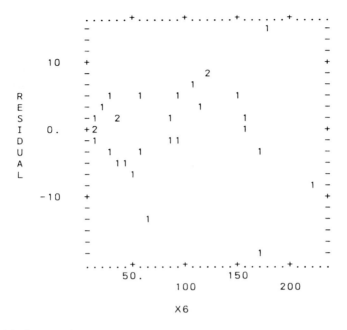

X6

From this figure it follows that the residuals are distributed fairly randomly about the line $e = 0$ so that the interaction term $X_1 X_5$ evidently does not need to be included in the model. Similar analyses can be done in respect of other interaction terms.

Finally, consider the identification of outliers. Although least squares residuals plots can be used for this purpose, studentized residuals which are discussed next, are more suitable for identifying possible outliers.

B. Studentized Residuals

The least squares residuals e_i defined previously generally have different variances and are also correlated. Studentized residuals are scaled least squares residuals that partially overcome some of the shortcomings accompanying least squares residuals and are defined as:

$$t_i = e_i/s_i^*, \qquad i = 1, 2, \ldots, n$$

where s_i^* represents the estimated standard deviation of e_i.

Weisberg (1980) points out that unlike the e_i, the sum of the t_i's is not zero and that the t_i are slightly correlated with Y although this correlation is usually neglible in practice. A major advantage of the studentized residuals is that they all have a variance equal to 1 as long as the model assumptions are met. Although the t_i do not exactly follow a t-distribution since the numerator and the denominator in the definition of t_i are not independent one may treat the t_i as approximately normally distributed for large samples. Cook and Weisberg (1982) distinguish between internal studentization of residuals (such as the t_i above) and external studentization of residuals in which the resulting residuals follow a t-distribution.

Internal studentized residuals can be obtained by means of the BMDP-9R program. Most of the residual plots presented in the case of least squares residuals can also be obtained in this case by appropriate specification in the PLOT paragraph. To obtain plots of studentized residuals against the predicted values, $\hat{Y}_i$ and against each of the predictors the following program could be used in which in the PLOT-paragraph the YVAR statement contains the 'Y' variable of the two-dimensional plot and the XVAR statement the 'X' variable of the two-dimensional plot.

Input: BMDP-9R program for obtaining studentized residual plots

```
/PROBLEM      TITLE IS 'REGRESSION OF Y ON X1-X5 '.
/INPUT        VARIABLES ARE 6.
              FORMAT = '(2F6.2,F3.0,2F4.0,F2.0)'.
/VARIABLE     NAMES ARE Y,X1,X2,X3,X4,X5.
/REGRESS      DEPENDENT=Y.
              INDEPENDENT=X1,X2,X3,X4,X5.
              METHOD=NONE.
/PRINT        MATR=RESI.
/PLOT         YVAR=STRESIDL,STRESIDL,STRESIDL,STRESIDL,STRESIDL,STRESIDL.
              XVAR=PREDICTD,X1,X2,X3,X4,X5.
              NORMAL.
              HISTOGRAM.
              SIZE=30,20.
/END
 34.50 32.00 12 214 173 5
 15.00 16.00  6 120  40 2
 29.00 27.00 10 156  36 1
               .
               .
               .

 13.80 13.50  5 112   0 4
 33.50 24.50  8 172  51 5
 10.90 10.50  4  77  20 5
/*
//
```

Output:

(i) *Plot of studentized residuals against* $\hat{Y}_i$

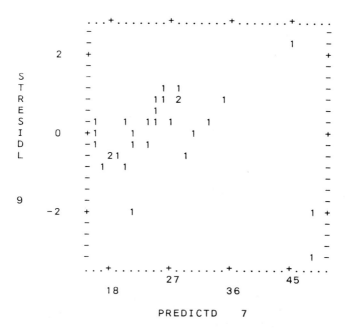

PREDICTD   7

Comparing the above plot with that of the least squares residuals against the $\hat{Y}_i$ it can be seen that, in this case, these two plots are nearly identical.

For this application the plots of the studentized residuals against the predictors are identical to those obtained for the least squares residuals against the predictors and are not presented here.

The need for including interaction terms can be investigated similarly to the case of least squares residuals where these residuals were plotted against the interaction term $X_1 X_5$. Adding the required TRANSFORM-paragraph the ADD and AFTER statements (as illustrated in the BMDP-1R program earlier), the YVAR and XVAR statements in the above BMDP-9R program would specify a plot of the studentized residuals (STRESIDL) against $X6$ $(X_1 X_5)$.

The NORMAL statement in the above BMDP-9R program produces a normal probability plot and since it is similar to that obtained for the least squares residuals by means of the BMDP-1R program, it is not given here.

A residual analysis can be useful in identifying outlier Y-values. Although the least squares residual plots are sometimes used in identifying such outliers the studentized residuals, because of their properties mentioned earlier, are more appropriate for this purpose. The t_i can for example, be represented by a histogram and since the t_i are approximately normally distributed under the assumptions of the regression model, Y values with absolute studentized residuals greater than 3 can be regarded as outliers. Such a histogram can be obtained by specifying the HISTOGRAM statement in the PLOT paragraph of the BMDP-9R program.

(ii) *Histogram of studentized residuals*

```
HISTOGRAM OF STANDARDIZED (STUDENTIZED) RESIDUALS
EACH BIN OF THE HISTOGRAM IS LABELED WITH ITS LOWER LIMIT.
NOTE THAT IF THE COUNT FOR A BIN EXCEEDS 100, ONLY
100 ASTERISKS WILL BE PRINTED.

          -3.4    1 *
          -3.2    0
          -3.0    0
          -2.8    0
          -2.6    0
          -2.4    0
          -2.2    0
          -2.0    2 **
          -1.8    0
          -1.6    0
          -1.4    0
          -1.2    0
          -1.0    1 *
          -0.8    3 ***
          -0.6    2 **
          -0.4    2 **
          -0.2    3 ***
           0.0    3 ***
           0.2    3 ***
           0.4    2 **
           0.6    2 **
           0.8    3 ***
           1.0    0
           1.2    2 **
           1.4    0
           1.6    0
           1.8    0
           2.0    0
           2.2    0
           2.4    1 *
```

Since one of the studentized residuals is smaller than -3 the corresponding Y-value could be regarded as an outlier. This Y-value can be identified by specifying the MATR = RESI option in a PRINT paragraph of the BMDP-9R program as shown in the input earlier.

The output obtained by specifying the MATR = RESI option identifies the 18th observation where $Y = 30.00$ as an outlier value.

C. Partial Residuals

Partial residuals are defined as:

$$r_i = Y_i - (\hat{Y}_i - \hat{\beta}_j X_{ij})$$
$$= e_i + \hat{\beta}_j X_{ij}.$$

Since

$$\hat{Y}_i = \hat{\alpha} + \hat{\beta}_1 X_{i1} + \hat{\beta}_2 X_{i2} + \cdots + \hat{\beta}_p X_{ip}, \qquad i = 1, 2, \ldots, n$$

it follows that $\hat{Y}_i - \hat{\beta}_j X_{ij}$ is an estimate of the i-th response when all the predictors except the jth one (X_j) are used. Hence the name partial residuals. A plot of r against the predictor X_j thus allows one to examine the relationship between Y and X_j after eliminating the influence of the other predictors.

Furthermore, in least squares residual plots of e against X_j the slope of the regression line of e against X_j can be expected to be zero whereas the regression of r against X_j should have a slope equal to $\hat{\beta}_j$, the coefficient of X_j when the full model was fitted. Gunst and Mason (1980) point out that it is this property of partial residuals which makes these plots useful in assessing the extent of possible nonlinearity in a certain predictor. If the slope of the plot of r against X_j approximately equals the coefficient $\hat{\beta}_j$ obtained from a fit of the full model the specification of X_j in the regression model can be assumed to be correct.

Plots of partial residuals r, against the predictors X_1, X_2, X_3, X_4 and X_5 for the data given in Table 7.1 can be obtained by means of the PREP $= X1, X2, X3, X4, X5$ statement in the BMDP-1R program given below.

Input: BMDP-1R program for obtaining partial residuals

```
/PROBLEM      TITLE IS 'REGRESSION OF Y ON X1-X5 '.
/INPUT        VARIABLES ARE 6.
              FORMAT = '(2F6.2,F3.0,2F4.0,F2.0)'.
/VARIABLE     NAMES ARE Y,X1,X2,X3,X4,X5.
/REGRESS      DEPENDENT=Y.
              INDEPENDENT=X1,X2,X3,X4,X5.
/PLOT         PREP=X1,X2,X3,X4,X5.
              SIZE=30,20.
/END
 34.50 32.00 12 214 173 5
 15.00 16.00  6 120  40 2
 29.00 27.00 10 156  36 1
              .
              .
              .
 13.80 13.50  5 112   0 4
 33.50 24.50  8 172  51 5
 10.90 10.50  4  77  20 5
/*
//
```

Output:

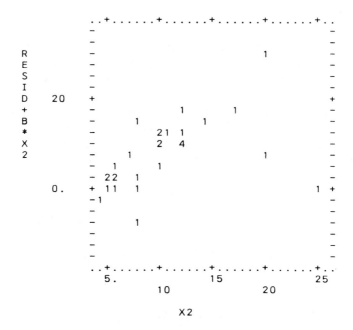

X2

Although the output contains partial residual plots for all the predictors only the plot of r against X_2 is given. From this plot, in which the effects of the predictors X_1, X_3, X_4 and X_5 have been removed, it appears that specifying X_2 in linear form in the regression equation is acceptable.

Gunst and Mason (1980, p. 250–251) provide an example where the plot of e against a predictor X_j seemingly indicates the need for an exponential transformation of that predictor. However, the plot of r against the predictor X_j indicates a proper specification of X_j in the model.

D. Predicted or Deleted Residuals

The residuals discussed thus far are all based on a fit in which all n observations are used. Predicted or deleted residuals, in contrast are obtained from a fit to the data in which the ith observation is deleted. Therefore, the ith deleted residual is defined as: $d_i = Y_i - \hat{Y}_{(i)}$, where $\hat{Y}_{(i)}$ is obtained from a fit in which the ith case was excluded. Mathematically it can be shown that the deleted residuals simply represent a scaling of the least squares residual. Similar to studentizing the least squares residual, deleted residuals can also be studentized to yield so-called studentized deleted residuals.

Deleted residual plots are of particular value in identifying outliers. Plots of the least squares residuals e, or the studentized residuals t, against the deleted residuals d should take a linear form if no outlier Y-values are present.

Such plots can be obtained by means of the YVAR = RESID, STRESIDL and XVAR = DELRESID, DELRESID statements in the BMDP-9R program given below.

Input: BMDP-9R program for obtaining predicted (deleted) residual plots

```
/PROBLEM      TITLE IS 'REGRESSION OF Y ON X1-X5 '.
/INPUT        VARIABLES ARE 6.
              FORMAT = '(2F6.2,F3.0,2F4.0,F2.0)'.
/VARIABLE     NAMES ARE Y,X1,X2,X3,X4,X5.
/REGRESS      DEPENDENT=Y.
              INDEPENDENT=X1,X2,X3,X4,X5.
              METHOD=NONE.
/PLOT         YVAR=RESIDUAL,STRESIDL.
              XVAR=DELRESID,DELRESID.
              NORMAL.
              HISTOGRAM.
              SIZE=30,20.
/END
 34.50 32.00 12 214 173 5
 15.00 16.00  6 120  40 2
 29.00 27.00 10 156  36 1
              .
              .
              .
 13.80 13.50  5 112   0 4
 33.50 24.50  8 172  51 5
 10.90 10.50  4  77  20 5
/*
//
```

Since the plots of *e* against *d*, and of *t* against *d* are very similar only the plot of the least squares residual *e* against the deleted residual *d* is given in the output below.

Output:

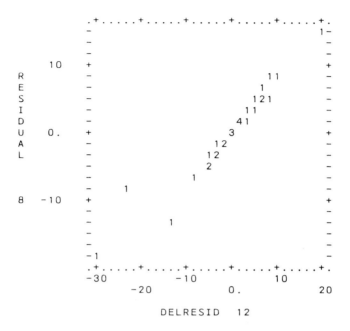

```
DELRESID   12
```

As was indicated by the histogram of studentized residuals the plot of *e* against the deleted residuals *d*, shows the existence of an outlier *Y*-value.

E. Recursive Residuals

While studentized residuals have equal variances, they are still correlated. Recursive residuals, on the other hand, are uncorrelated and as pointed out by Galpin and Hawkins (1984), may be interpreted as showing the effect of successively deleting observations from the data set. These authors illustrate the use of these residuals to check the assumptions of normality, homoscedasticity of the errors, the existence of outliers and the omission of predictors.

Since recursive residuals follow independent identical $n(0, \sigma^2)$ distributions under the full model, they have the further advantage that exact tests may be applied to them. These residuals are particularly suited to data which can be ordered in some natural way but Galpin and Hawkins (1984), however, maintain that recursive residuals are useful for all data sets.

Let $\hat{\boldsymbol{\beta}}_r$ be the least squares regression vector based on the first r observations. Further, let the $n \times 1$ response vector, $\mathbf{Y}$ and the $n \times (p + 1)$ matrix $\mathbf{X}$ consisting of the n observations on the $p + 1$ predictors (the intercept included) be partitioned accordingly. The relevant matrix components from which $\hat{\boldsymbol{\beta}}_r$ is calculated are:

$$\tilde{\mathbf{X}}_r' = (\mathbf{X}_1, \mathbf{X}_2, \ldots, \mathbf{X}_r) \quad \text{and} \quad \mathbf{Y}_r' = (Y_1, Y_2, \ldots, Y_r).$$

If $\tilde{\mathbf{X}}_r'\tilde{\mathbf{X}}_r$ is nonsingular then $\hat{\boldsymbol{\beta}}_r = (\tilde{\mathbf{X}}_r'\tilde{\mathbf{X}}_r)^{-1}\tilde{\mathbf{X}}_r'\mathbf{Y}_r$ and the recursive residuals are defined as:

$$w_r = \frac{Y_r - \mathbf{X}_r'\hat{\boldsymbol{\beta}}_{r-1}}{\sqrt{1 + \mathbf{X}_r'(\tilde{\mathbf{X}}_{r-1}'\tilde{\mathbf{X}}_{r-1})^{-1}\mathbf{X}_r}}.$$

Obviously only $n - (p + 1)$ recursive residuals can be calculated since at least $(p + 1)$ points are needed for the fitting of a $(p + 1)$-parameter regression model. The name recursive residual derives from the fact that w_r can be obtained from w_{r-1} by means of an updating formula.

When calculating recursive residuals a so-called base set of $(p + 1)$ observations is chosen on which the initial regression is based. Great care needs to be exercised that this base set does not pose problems of singularity in the resulting regression run.

On the assumptions that the errors are identically and independently distributed as $n(0; \sigma^2)$, the recursive residuals are also independently and identically distributed as $n(0; \sigma^2)$ so that if the model assumptions are satisfied, the normal probability plot should show a straight line through the origin.

Galpin and Hawkins (1984) illustrate the use of plots of the CUSUMS of the recursive residuals and plots of the CUSUMS of the square roots of the absolute values of the standardised recursive residuals to check the model assumptions mentioned earlier, to investigate the effect of an omitted variable and to detect outliers. Although CUSUMS are discussed only in Chapter 9 an illustration of a CUSUM plot of the recursive residuals to identify outliers, will be briefly discussed next.

The ith CUSUM of the recursive residuals is defined as:

$$\text{CUSUM}_i = \sum_i w_i, \qquad i = 1, 2, \ldots, n - (p + 1).$$

The CUSUM plot considered here is a plot of CUSUM_i/s against i where $s^2 = \sum_i w_i^2/(n - p - 1)$. If the outliers are among the calculated recursive residuals these authors point out that the plot of the CUSUM against the CUSUM-number will show a sudden blip (corresponding to the large recursive residual), preceded by a downward drift. If the outlier is among the base set of observations, the plot will show a nonhorizontal straight line indicating that the mean of the recursive residuals is nonzero. The two cases described here respectively correspond to normal probability plots which show a straight line through the origin with a few points off the line and a plot that does not pass through the origin.

Recursive residuals for a regression analysis on the data set given in Table 7.1 were calculated by means of the following SAS PROC MATRIX program. A normal probability plot of the w_r and a plot of the CUSUM against the CUSUM-number was obtained by means of the accompanying SAS PROC RANK and SAS PROC PLOT programs.

Input: SAS PROC MATRIX and SAS PROC PLOT programs for calculating and plotting recursive residuals

```
DATA    GENERAL ;
INPUT   NUMBER Y X1-X5 ;
CARDS ;
    1    34.50 32.00 12 214 173 5
   11    60.00 60.00 20 448  62 3
   12    40.00 38.00 17 402   0 4
   13    40.00 73.50 20 523   0 3
    2    15.00 16.00  6 120  40 2
    3    29.00 27.00 10 156  36 1
    4    30.00 29.00 10 280  42 4
    5    29.00 24.50 12 183  62 1
    6    33.00 30.00 11 136  47 3
    7    25.60 22.00 10 228   0 4
    8    15.00 17.00  8 170  42 2
    9    17.50 13.50  6 180   0 1
   10    33.50 28.00 14 159  46 2
   14    15.00 14.00  6 111 107 3
   15    27.00 34.00  8 198 124 5
   16    35.99 23.99 12 194  25 5
   17    20.50 18.50  6 115  34 5
   18    30.00 57.00 25 406  71 3
   19    13.80 12.50  5 133  37 1
   20    21.00 16.50 10 134   0 5
   21    25.00 19.50 12 127  36 2
   22    21.50 15.00 12  98  14 1
   23    32.00 22.00 10 151  45 5
   24     8.90 16.00  8 142  54 4
   25    30.00 31.00 12 237  36 5
   26    22.40 17.90  7 150 102 2
   27    14.90 12.00  5 140  35 1
   28    13.80 13.50  5 112   0 4
   29    33.50 24.50  8 172  51 5
   30    10.90 10.50  4  77  20 5
;
PROC    MATRIX ;
FETCH   XX ;
        N= NROW(XX) ;
        NCLX = NCOL(XX) ;
        Y =  XX(,2) ;                        *  Dependant variable ;
        X = J(N,1,1)||XX(,3:NCLX) ;          *  Design matrix ;
        K = NCOL(X)+1 ;
        SEQ= XX(K:N,1) ;
        FREE XX ;
        NPAR = K-1 ;                         * Number of parameters ;
        SR = J(N-NPAR,1,0) ;                 * Initialize ;
        WR = J(N-NPAR,1,0) ;
        CUSUM = J(N-NPAR,1,0) ;
        XR = X(1:NPAR, ) ;                   * Obtain base set ;
        YR = Y(1:NPAR,1) ;
        TEMPR = INV(XR'*XR) ;
        BR = TEMPR*XR'*YR ;                  * B-hat based on npar values ;
        RESR = YR - XR*BR ;
        SRR = RESR'*RESR ;                   * Sum of residuals_squared ;
DO R = K TO N ;                              * Calculate recursive residuals ;
        VXR = X(R,  )' ;
        WR(R-K+1,1) = (Y(R,1) - VXR'*BR)#/SQRT(1 + VXR'*TEMPR*VXR) ;
        D = TEMPR*VXR ;
        TEMPR = TEMPR- ((D*D')#/(1+VXR'*D))  ;
        BR = BR+ TEMPR*VXR*(Y(R,1) - VXR'*BR) ;
        IF R=K THEN SR(R-K+1,1) = SRR   + WR(R-K+1,1)**2 ;
        IF R>K THEN SR(R-K+1,1) = SR(R-K,1) + WR(R-K+1,1)**2 ;
END ;                                        * End of K_loop ;
        STD = SQRT(SR(N-NPAR,1)#/(N-NPAR)) ;
        SUMRES = 0.0 ;
        NMNP = N - NPAR ;
```

```
DO I = 1 TO NMNP ;                        * Calculate CUSUM ;
      SUMRES = SUMRES + WR(I,1) ;
      CUSUM(I,1) = SUMRES#/STD ;
END ;                                     * End of I-loop ;
SUMM = SEQ||WR||SR||CUSUM ;               * Summary Statistics ;
OUTPUT SUMM OUT = WRSR (RENAME = (COL1 = NUMB COL2 = RECRES COL3 = SSQ
                        COL4 = CUSUM )) ;
PROC    PRINT ;
PROC    PLOT DATA = WRSR ;
PLOT    CUSUM*NUMB = 'x' / VPOS = 30 HPOS = 50 ;
PROC    RANK DATA = WRSR NORMAL=BLOM OUT = NORMPLOT ;
        VAR RECRES ;
        RANKS RANKRES ;
PROC    PRINT DATA = NORMPLOT ;
PROC    PLOT DATA = NORMPLOT ;
PLOT    RANKRES*RECRES = '*' / VPOS= 30   HPOS= 50 ;
```

Output:

(i) *Recursive residuals (RECRES), residual sum of squares (SSQ) and CUSUM values*

NUMB	RECRES	SSQ	CUSUM
4	5.193	26.97	0.7305
5	-11.614	161.85	-0.9032
6	5.133	188.20	-0.1812
7	-0.016	188.20	-0.1835
8	-5.180	215.03	-0.9121
9	5.524	245.54	-0.1350
10	-3.934	261.02	-0.6884
14	-3.773	275.26	-1.2192
15	2.421	281.12	-0.8786
16	2.970	289.95	-0.4607
17	0.827	290.63	-0.3444
18	-23.361	836.37	-3.6305
19	-2.559	842.92	-3.9904
20	-5.046	868.39	-4.7003
21	0.524	868.66	-4.6266
22	-0.483	868.90	-4.6946
23	4.651	890.53	-4.0403
24	-13.651	1076.88	-5.9606
25	-0.733	1077.42	-6.0636
26	2.332	1082.86	-5.7356
27	-0.804	1083.51	-5.8487
28	-4.669	1105.31	-6.5055
29	7.798	1166.11	-5.4086
30	-6.843	1212.93	-6.3712

(ii) *Normal probability plot of the recursive residuals*

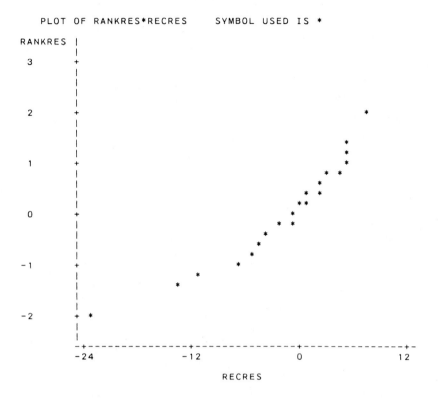

PLOT OF RANKRES*RECRES   SYMBOL USED IS *

NOTE: 3 OBS HIDDEN

(iii) *Plot of CUSUM of the recursive residuals against the CUSUM-number*

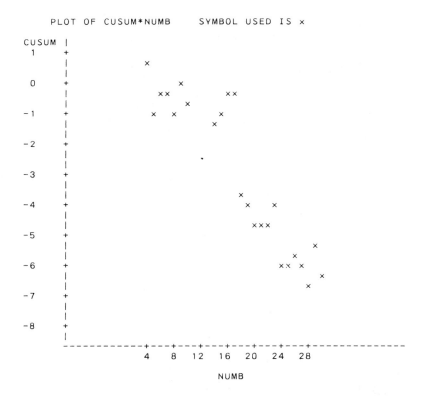

Upon comparing the above normal probability plot with the corresponding normal probability plot of the least squares residuals the same heavy tailed distribution, indicating deviations from normality, is present. Furthermore, the CUSUM-plot shows a clear blip for $i = 18$ which corresponds to the identification of the $Y = 30.00$ observation as an outlier as was found earlier by means of the studentized residual plot.

4. Mallows' C_k-Statistic

A particularly important aspect of a regression analysis with a large number of explanatory variables is the choice of an appropriate subset of predictors which should be included in the model. In a regression analysis an attempt is usually made to obtain the simplest possible model that satisfactorily explains the underlying structure of the data. Various analytical techniques for obtaining such a subset of the explanatory variables exist (for instance stepwise regression procedures). A graphical technique, the representation of Mallows' C_k-statistic, will now be discussed in this connection. For each subset of k

predictors (including the intercept) the C_k-statistic can be calculated as:

$$C_k = (ESS_k/s^2) - (n - 2k),$$

where $ESS_k = \sum e_i^2$ is the error sum of squares that is obtained by fitting a model with k parameters (including the intercept term) and s^2 is an estimate of σ^2 which is obtained by including all p explanatory variables in the regression model.

If the $(p + 1) - k$ variables not included in the regression model do not contribute to the explaining of the dependent variable, it can be shown that C_k will approximately equal k. In a regression analysis the value of C_k is calculated for the different possible subsets of explanatory variables and C_k-values are then plotted against k. Sets of variables which correspond to points close to the line $C_k = k$ represent the sets from which the eventual regression models can be chosen. Mallows (1973) suggests that good models will have negative or small $C_k - k$ values.

For illustrative purposes Mallows C_k (or C_p statistic as it is also known) is calculated for the data of Table 7.1 by means of the following BMDP-9R program.

Input: BMDP-9R program for Mallows' C_k-statistic

```
/PROBLEM      TITLE IS 'REGRESSION OF Y ON X1-X5 '.
/INPUT        VARIABLES ARE 6.
              FORMAT = '(2F6.2,F3.0,2F4.0,F2.0)'.
/VARIABLE     NAMES ARE Y,X1,X2,X3,X4,X5.
/REGRESS      DEPENDENT=Y.
              INDEPENDENT=X1,X2,X3,X4,X5.
              METHOD=CP.
/END
 34.50 32.00 12 214 173 5
 15.00 16.00  6 120  40 2
 29.00 27.00 10 156  36 1
               .
               .
               .
 13.80 13.50  5 112   0 4
 33.50 24.50  8 172  51 5
 10.90 10.50  4  77  20 5
/*
//
```

From the output the C_k-plot given in Figure 7.3 was obtained.

Subsets of predictors in the area indicated by the dotted line could be considered as plausible subsets of predictors. "Best" subsets of predictors could be chosen by considering amongst other things the so-called adjusted R^2-value which is discussed in, for example, Weisberg (1980), p. 188.

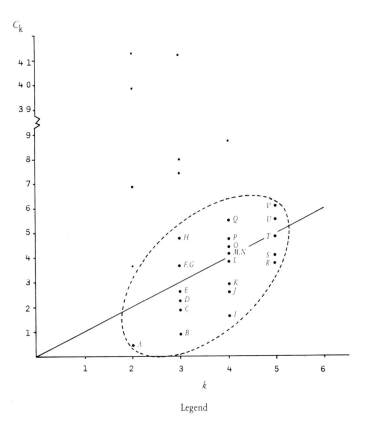

Legend

$A = \{X_1\}$ $B = \{X_1, X_2\}$ $I = \{X_1, X_2, X_5\}$ $R = \{X_1, X_2, X_4, X_5\}$

$C = \{X_1, X_5\}$ $J = \{X_1, X_2, X_4\}$ $S = \{X_1, X_2, X_3, X_5\}$

$D = \{X_1, X_4\}$ $K = \{X_1, X_2, X_3\}$ $T = \{X_1, X_2, X_3, X_4\}$

$E = \{X_1, X_3\}$ $L = \{X_1, X_4, X_5\}$ $U = \{X_2, X_3, X_4, X_5\}$

$F = \{X_2, X_3\}$ $M = \{X_1, X_3, X_5\}$ $V = \{X_1, X_3, X_4, X_5\}$

$G = \{X_2, X_5\}$ $N = \{X_2, X_3, X_5\}$

$H = \{X_2, X_4\}$ $O = \{X_1, X_3, X_4\}$

$P = \{X_2, X_3, X_4\}$

$Q = \{X_2, X_4, X_5\}$

Figure 7.3. Mallows' C_k-plot.

5. Confidence and Forecast Bands

In the case of $p = 1$ the 95% confidence interval for the mean value of Y for a specific value of X is given by

$$\hat{\alpha} + \hat{\beta}X \pm t_{n-2;0.025}s\sqrt{1/n + (X - \bar{X})^2/(\sum X_i^2 - n\bar{X}^2)}.$$

The graphical representation of these confidence intervals for different values of X leads to the so-called confidence band for the population regression line or the mean value of Y.

The forecast band for the individual Y-values for a specified value of X is obtained from the representation of the 95% confidence intervals for the individual Y-values given by

$$\hat{\alpha} + \hat{\beta}X \pm t_{n-2;0.025}s\sqrt{1 + 1/n + (X - \bar{X})^2/(\sum X_i^2 - n\bar{X}^2)}.$$

The representation of these confidence bands can be obtained by means of SAS/GRAPH and will now be illustrated by way of an example.

For the data given in Table 7.1 consider the case $Y = \alpha + \beta X_1 + \varepsilon$.

The GPLOT-section of the SAS/GRAPH program which was used to obtain the 95% confidence band for the mean value of Y by utilizing the I-option of the SYMBOL statement is shown below.

Input: SAS/GRAPH PROC GPLOT program for the representation of a 95% confidence band

```
DATA    HOUSES ;
INPUT   Y X1 @@ ;
CARDS ;
 34.50 32.00   15.00 16.00   29.00 27.00   30.00 29.00   29.00 24.50
 33.00 30.00   25.60 22.00   15.00 17.00   17.50 13.50   33.50 28.00
 60.00 60.00   40.00 38.00   40.00 73.50   15.00 14.00   27.00 34.00
 35.99 23.99   20.50 18.50   30.00 57.00   13.80 12.50   21.00 16.50
 25.00 19.50   21.50 15.00   32.00 22.00    8.90 16.00   30.00 31.00
 22.40 17.90   14.90 12.00   13.80 13.50   33.50 24.50   10.90 10.50
;
FOOTNOTE1 .C=BLACK 'Regression of Y on X1' ;
FOOTNOTE2 .C=BLACK '95 Percent confidence band' ;
PROC    GPLOT ;
PLOT    Y*X1 / HAXIS = 0 TO 80 BY 10 VAXIS = 0 TO 60 BY 10 ;
SYMBOL1 I=RLCLM95 ;
```

Output:

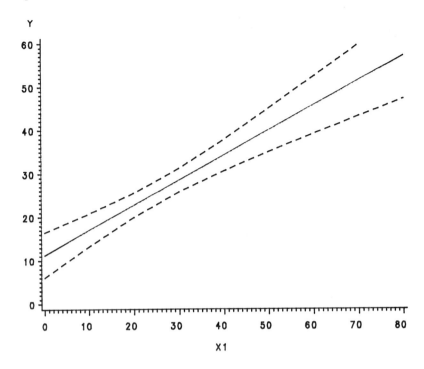

Regression of Y on X1
95 Percent confidence band

Suppose the mean value of Y must be estimated for $X_1 = 40$. The above figure shows a 95% confidence interval of $(30, 38)$. The rapidly broadening confidence band for larger X_1-values shows that extrapolation in this case would not be very informative.

A forecast band is obtained by replacing CLM with CLI in the I option of the SYMBOL statement.

6. The Ridge Trace

A meaningful regression analysis requires a high correlation of each of the explanatory variables $X_1, X_2, \ldots, X_p$ with the dependent variable, while at the same time having a low correlation with each of the other explanatory variables. Explanatory variables that are nearly linearly dependent are termed collinear. In this case the problem of multicollinearity arises.

Pairwise collinearity can be investigated by the matrix of Pearson correlation coefficients. However, the identification of multicollinearities between three or more explanatory variables does not simply follow from an analysis of the pairwise correlations but usually requires the calculation of more advanced statistical measures such as variance inflation factors and condition indices described in, for example, Belsley, Kuh, and Welsch (1980). These measures can be obtained by, for example, the SAS PROC REG program.

The most important consequences of a least squares regression analysis in the case of multicollinearities being present, are as follows:

(i)   The variances of the estimated regression coefficients can become very large with the result that these estimates can be regarded as inaccurate.
(ii) Different stepwise regression methods can result in contradictory regression equations.
(iii) The signs of the estimated regression coefficients can be wrong.
(iv) Prediction error in extrapolating can be severe.

To counter the problem of multicollinearity, estimators of the regression coefficients can be developed which may be somewhat biased but which also have considerably smaller variances than the ordinary least squares estimators. In ridge regression for example, the ordinary least squares estimators are adjusted, thereby reducing the influence of multicollinearities. Ridge estimators are defined as a function of a ridge parameter λ. A discerning choice of λ can ensure that the ridge estimator of the regression coefficients is more accurate than the least squares estimator.

A plot of the so-called ridge trace yields a technique for obtaining a suitable value of λ. Ridge estimators of the $\beta_j, j = 1, 2, \ldots, p$ are calculated for different values of λ in the interval $(0, 1)$ and then plotted against λ. A value for λ is then chosen in the range of values where the estimates of the regression coefficients begin to stabilize.

Since interpreting a ridge-trace plot involves direct comparison of the $\hat{\beta}_j$, the ridge trace is usually obtained from a regression analysis in which all the variables are standardized. Such a standardization neutralises the effect which units of measurement of the predictor X_j have on $\hat{\beta}_j$. This standardization can be carried out by means of the SAS PROC STANDARD program.

Ridge regression can be carried out by means of the BMDP-2R program as described in Appendix C.7 of the BMDP (1983) manual. This program however, does not plot the ridge trace. Ridge estimates for various values of λ in a regression analysis can be calculated by means of the following SAS PROC MATRIX program and the subsequent ridge trace can be obtained by means of of the SAS/GRAPH PROC GPLOT program.

To illustrate a regression in which multicollinearities occur consider a response surface model of Y with $X_1, X_2, X_3, X_1^2, X_1 X_2, X_2^2, X_3^2, X_2 X_3$ and $X_1 X_3$ as predictors for the data given in Table 7.1.

Input: SAS PROC MATRIX program and SAS/GRAPH GPL OT program for obtaining ridge estimates and for plotting the ridge trace

```
DATA    HOUSES ;
*       Y = LOAN AMOUNT GRANTED ( IN R1000 ) ;
INPUT   Y X1- X3 ;
        X4=X1*X1 ;
        X5=X1*X2 ;
        X6=X1*X3 ;
        X7=X2*X2 ;
        X8=X2*X3 ;
        X9=X3*X3 ;
CARDS ;
34.50 32.00  12 214
15.00 16.00   6 120
29.00 27.00  10 156
            .
            .
            .
13.80 13.50   5 112
33.50 24.50   8 172
10.90 10.50   4  77
;
PROC    STANDARD OUT=A MEAN=0 STD=1 ;
PROC    MATRIX ;
        FETCH Z DATA = A ;          * STANDARDIZED OBSERVATIONS ;
        N=NROW(Z) ;
        NC=NCOL(Z) ;
        Y=Z(,1) ;
        X=Z(,2 : NC ) ;             * DESIGN MATRIX ;
        XTX=X'*X ;
        P=NROW(XTX) ;
        ID=I(P) ;
DO I=0 TO 27 ;
        L=0.001*(1.20**(I-1)) ;  * GEOMETRIC INCREASE IN L ;
        IF I=0 THEN L=0 ;
        LAMB=L*ID ;
        BETA=INV(XTX+LAMB)*X'*Y ;
        TEMP=L||BETA' ;
        OUTPUT TEMP OUT=B (RENAME=(COL1=L COL2=B1 COL3=B2 COL4=B3
        COL5=B4 COL6=B5 COL7=B6 COL8=B7 COL9=B8 COL10=B9)) ;
END ;
GOPTIONS COLORS=(BLACK) PENMOUNTS=1 ;
PROC    PRINT DATA=B ;
PROC    GPLOT ;
        PLOT (B1-B9)*L=1  / OVERLAY VREF=0 HAXIS = 0 TO 0.12 BY 0.02 ;
        SYMBOL1 I=JOIN L=1 ;
FOOTNOTE 'Ridge Trace' ;
```

Output:

(i) *Table of parameter estimates for values of λ*

L	B1	B2	B3	B4	B5	B6	B7	B8	B9
0.000000	1.95819	1.56617	-1.0571	-9.8614	8.68262	6.26307	-3.9874	-3.7093	0.629737
0.001000	1.82530	1.60004	-0.9767	-7.9234	7.32721	4.38917	-3.9483	-2.4091	0.627413
0.001200	1.80772	1.60296	-0.9642	-7.6395	7.11537	4.12480	-3.9335	-2.2212	0.623383
0.001440	1.78925	1.60537	-0.9501	-7.3303	6.87984	3.84048	-3.9138	-2.0177	0.617801
0.001728	1.77025	1.60696	-0.9346	-6.9978	6.62056	3.53939	-3.8882	-1.8005	0.610371
0.002074	1.75119	1.60737	-0.9175	-6.6451	6.33811	3.22579	-3.8553	-1.5727	0.600832
0.002488	1.73257	1.60620	-0.8988	-6.2764	6.03374	2.90490	-3.8138	-1.3377	0.589005
0.002986	1.71494	1.60303	-0.8786	-5.8966	5.70942	2.58260	-3.7622	-1.1000	0.574837
0.003583	1.69884	1.59745	-0.8568	-5.5111	5.36775	2.26505	-3.6994	-0.8646	0.558446
0.004300	1.68474	1.58907	-0.8333	-5.1255	5.01193	1.95823	-3.6240	-0.6366	0.540149
0.005160	1.67298	1.57760	-0.8082	-4.7450	4.64558	1.66750	-3.5354	-0.4212	0.520457
0.006192	1.66375	1.56283	-0.7811	-4.3745	4.27270	1.39727	-3.4331	-0.2231	0.500049
0.007430	1.65701	1.54470	-0.7522	-4.0182	3.89745	1.15073	-3.3171	-0.0464	0.479695
0.008916	1.65254	1.52329	-0.7211	-3.6790	3.52408	0.92972	-3.1882	0.1058	0.460159
0.010699	1.64987	1.49884	-0.6876	-3.3593	3.15676	0.73487	-3.0476	0.2314	0.442084
0.012839	1.64835	1.47176	-0.6517	-3.0602	2.79947	0.56567	-2.8971	0.3296	0.425883
0.015407	1.64718	1.44254	-0.6131	-2.7822	2.45591	0.42078	-2.7389	0.4006	0.411666
0.018488	1.64546	1.41180	-0.5716	-2.5253	2.12937	0.29828	-2.5758	0.4457	0.399206
0.022186	1.64226	1.38019	-0.5272	-2.2886	1.82262	0.19590	-2.4105	0.4673	0.387963
0.026623	1.63667	1.34836	-0.4797	-2.0713	1.53790	0.11128	-2.2459	0.4680	0.377160
0.031948	1.62790	1.31689	-0.4294	-1.8721	1.27679	0.04208	-2.0847	0.4513	0.365893
0.038338	1.61533	1.28627	-0.3764	-1.6897	1.04025	-0.01390	-1.9293	0.4204	0.353261
0.046005	1.59852	1.25683	-0.3209	-1.5229	0.82858	-0.05863	-1.7814	0.3787	0.338485
0.055206	1.57728	1.22877	-0.2634	-1.3705	0.64150	-0.09389	-1.6426	0.3294	0.321006
0.066247	1.55163	1.20210	-0.2047	-1.2311	0.47820	-0.12124	-1.5137	0.2753	0.300550
0.079497	1.52183	1.17669	-0.1453	-1.1039	0.33745	-0.14201	-1.3950	0.2189	0.277147
0.095396	1.48827	1.15225	-0.0862	-0.9880	0.21769	-0.15737	-1.2864	0.1622	0.251110
0.114475	1.45147	1.12841	-0.0283	-0.8825	0.11718	-0.16829	-1.1876	0.1071	0.222989

(ii) *Plot of ridge trace*

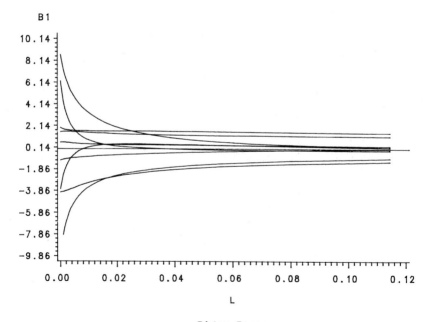

Ridge Trace

From this output it follows that the estimates of the β_j begin to stabilize in the range of λ values for which $0.02 < \lambda < 0.04$ so that a possible choice of λ would be 0.03. From the table in the above output for $\lambda = 0.03$ more suitable estimates of the regression coefficients can now be obtained.

7. General

In recent years a number of residuals have been developed for use in regression analysis. These residuals, illustrated by examples, are discussed extensively by, among others, Cook and Weisberg (1982), Draper and Smith (1981), Galpin and Hawkins (1984) and Gunst and Mason (1980).

From the above discussion on graphical representations in regression analysis it is evident that residual plots are a particularly useful and instructive way of evaluating the usual assumptions of a regression analysis. In representing residuals the greatest shortcoming is most probably the degree of subjectivity that could arise in the evaluation of the plots. It is therefore essential to gain wide experience in residual analysis, and continual reference, particularly to the above-mentioned sources is strongly recommended.

CHAPTER 8

CHAID and XAID: Exploratory Techniques for Analyzing Extensive Data Sets

1. Introduction

In many situations techniques are called for which will enable the researcher to identify particular patterns in the data and which can be used for formulating any structural relationships between the variables. The computer programs CHAID and XAID, which are to be discussed in this chapter, are both examples of so-called AID procedures (Automatic Interaction Detection), according to which the outcome of a dependent variable Y can be predicted on the basis of those predictors (independent variables) which contribute the most to the variation in Y.

The computer program CHAID was developed for analyzing categorical data, which implies that both the dependent and independent variables are categorical in nature. In the case of XAID however, a continuous dependent variable is used.

2. CHAID—An Exploratory Technique for Analyzing Categorical Data

The computer program CHAID originally developed by Kass (1980) is a procedure for predicting the outcome of a categorical dependent variable Y on the basis of a set of independent categorical variables. In this way predictors which play a role are identified and the data can be analyzed by means of a relevant technique such as the logit form of the log-linear model or an applicable regression analysis technique. Interactions between predictors can also be detected by using the program CHAID. In analyzing large data sets

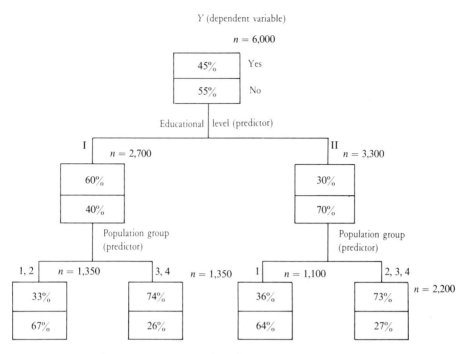

Figure 8.1. Representation of a dendrogram in CHAID.

the CHAID program can be used as an exploratory technique aimed at reducing the dimensions of the relevant contingency table.

If a researcher wants to predict the outcome of a dependent variable Y on the basis of p predictors, CHAID's computer output is used for compiling a dendrogram. Figure 8.1 shows such a typical dendrogram. This figure is based on the response of 6000 persons on a dichotomous dependent variable, where inter alia educational level and population group were used as predictors.

From this figure it follows that educational level as predictor explains more of the variation in Y than any other predictor included in the analysis. This predictor is divided into two categories (I and II). The second most important predictor is population group, which has four categories. At Educational Level I, population group is divided into two classes, namely 1, 2 and 3, 4 whereas at Educational Level II, the subdivision is into classes 1 and 2, 3, 4. In the case of Educational Level II for instance, 1 and 2, 3, 4 can be regarded as two classes which statistically differ significantly in respect of the way in which each explains the variation in Y. Each class however, forms a more or less homogeneous group. In the above example, further predictors such as sex, language and age were also included in the CHAID-analysis. However, since their contribution is not statistically significant in respect of question Y, they do not form part of the dendrogram. The program CHAID is used mainly as:

(a) An exploratory analysis for determining a relevant model to which further statistical techniques can be applied.
(b) A final complete analysis. Thus a Y value can be predicted which corresponds to a particular level of a predictor. This is done by following the branchings of the dendrogram from Y to the required level of the particular predictor. From the figure it follows, for instance, that 33% of the population who are members of Population Groups 1 and 2 and Educational Level I, are expected to answer "yes" to question Y whereas 67% are expected to answer "no".

In a CHAID analysis it is assumed that the predictors as well as the dependent variable are categorical variables (i.e., nominal or ordinal). Nominal variables include, for example, sex (male, female), language (English, French, German) and marital status (married, divorced, unmarried). An example of an ordinal variable is income (low, average, high). A predictor such as age which can be measured on an interval scale can be transformed into an ordinal variable for the purpose of using CHAID by placing the possible outcomes of the variable in categories. Thus the variable age can be categorized as (18 to 30), (31 to 49) and (50 to 64), i.e. three categories of which the levels are 1, 2 and 3 respectively.

As part of the CHAID input it should be indicated whether the predictors are monotone or free. For the purpose of CHAID a predictor is defined as monotone if the relationship between such a predictor and the dependent variable Y is in fact monotone in nature, i.e. if the values of Y increase (or decrease) only in accordance with an increase in the values of the predictor.

All ordinal variables are therefore not monotone in nature. Thus the relationship between the ordinal variables age (as predictor) and income (as dependent variable) may first show an ascending and then a descending trend. In this case age is regarded as a "free" variable in a CHAID analysis. Usually predictors such as residential area, language, population group and marital status are also regarded as "free" variables.

For a given data set a CHAID analysis involves a number of steps, each of which has the following form:

(a) Stratification of each predictor in respect of the dependent variable Y. This occurs because the categories (levels) of a particular predictor are checked and possibly regrouped into a number of classes, say k, each of which is more or less homogeneous in respect of the Y values. Suppose the six categories of a predictor are indicated by the symbols A, B, C, D, E and F. These categories are for instance then reduced to three classes, namely $\{A, D\}$, $\{B\}$ and $\{C, E, F\}$ and in such a way that there are significant differences between the three classes in respect of their explanation of the variation of Y, but not between the categories in each class.
(b) After each of the predictors has been analyzed in the above manner, the predictor which explains most of the variation in the Y values is used to divide the data into k subsets.

From Figure 8.1, for instance, it follows that the predictor, educational level, was divided into two separate classes in Step (a) and in Step (b) was chosen as the predictor which explains most of the variation in the observed Y values.

Each subset is subsequently analyzed according to Steps (a) and (b) described above. The process is continued until no further statistically significant division of the data into subsets is possible.

The statistical criteria used in CHAID for subdividing data into subdata sets are:

(i) The χ^2-statistic used in $r \times c$ contingency tables, where r indicates the number of levels of Y and c the number of levels of the predictor under discussion.

(ii) Bonferroni intervals for the level of significance of the test. Suppose the original c categories of a predictor can be subdivided into k classes in B different ways, then a $100(1 - \alpha)\%$ Bonferroni interval is calculated by determining the critical values of the particular test statistic at a α/B level of significance.

Interpreting the Dendrogram

As already mentioned, the computer program CHAID's output can be used for compiling a dendrogram. A further example of such a diagram is given in Figure 8.2. The dependent variable Y in this case is: Number of days absent from work (in days per year). The Y values are grouped into the categories: "seven days and fewer" and "more than seven days". The variables age, sex, level of skill and geographical area were used as predictors. Only two of these predictors, namely sex and level of skill appear in the dendrogram, which means that age and geographical area may be ignored in further model formulation. From Figure 8.2 it appears that sex is the most important predictor. Both subgroups, male and female, are divided into two subgroups of level of skill. The following information is obtained from the dendrogram.

(i) *The Final Groupings.* The workers in the sample can be subdivided into four groups, namely (a) skilled and semi-skilled male workers, (b) unskilled male workers, (c) skilled and semi-skilled female workers, and (d) unskilled female workers. These subgroups form the basis for a model according to which absence from work can be predicted.

(ii) *Association.* There is a definite relationship between the predictors sex and level of skill. The ratio of unskilled female workers (500/700) is considerably larger than the ratio of unskilled male workers (350/800) and this partly explains why there is a statistically significant difference between the absence from work of men and women.

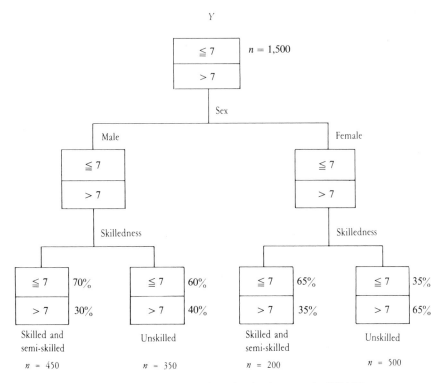

Figure 8.2. Representation of a dendrogram in CHAID.

(iii) *Interaction.* Suppose a predictor A divides the data set into subgroups. Furthermore, suppose that at the second level of partitioning each subgroup is similarly divided into further subgroups on the basis of predictor B. Then the model is additive in A and B. An AB interaction is present however, if the partitioning of subgroups at the second partitioning level occurs in different ways, or if the partitioning applies only to some of the subgroups formed by A. An ABC interaction is indicated similarly by an unbalanced structure of the dendrogram with regard to the predictors A, B and C.

The presence of interaction between sex and level of skill is indicated by the difference in the effect of level of skill on men and women. Note for instance that in the case of men there is a 10% difference in work absence between the two subgroups as opposed to a 30% difference for women. In addition the sex effect for unskilled workers is 25% as against only 5% for more skilled workers.

CHAID is not reliable for use with small data sets and progressively provides more meaningful results as the number of observations increase. A general guideline is to ensure that every subgroup or class consists of a minimum of $5 \times r \times c$ observations, where r is the number of levels of Y (dependent variable) and c the number of levels of the predictor. A more general guideline is to use only data sets with 500 or more observations in a

CHAID analysis. A further restriction concerns the number of levels or categories of the predictors which should not exceed 10.

3. Applying a CHAID Analysis

The full Ability Test Data set ($n = 2747$) described in Chapter 1 was used in order to illustrate the application of CHAID. As dependent variable Y, proposed university study field (X_{21}) was used, which was divided into four categories as follows:

Category 1: Medicine, engineering and natural sciences degree.
Category 2: Arts, law, commerce and administration degree.
Category 3: Medical or education diploma.
Category 4: No further university study.

The eight predictors used are language group (two categories), sex (two categories), numbers series (X_{13}), figure analogies (X_{14}), pattern completion (X_{15}), word pairs (X_{16}), verbal reasoning (X_{17}) and word analogies (X_{18}). The variables X_{13} to X_{18} all refer to Educational Level III and were transformed into ordinal variables by calculating quartiles (q). Variable X_i ($i = 13, 14, \ldots, 18$) is therefore categorized as follows:

Category 1: $X_i < q_1$
Category 2: $q_1 \leq X_i < q_2$
Category 3: $q_2 \leq X_i < q_3$
Category 4: $X_i \geq q_3$

The CHAID computer program requires as input:

(i) a data file containing particulars on the variables as well as on program options, and
(ii) a data file containing the input format as well as the observed data.

The SAS program facilities can be used for creating both these files.
 The input data are as follows:

Input: CHAID program for the exploratary analysis of categorical data

(i) *Particulars of variables and program options*

```
                                                              Section
     'STUDY FIELD'  '4'                                          1
     '1'  '2'  '3'  '4'                                          2
      8                                                          3
     'LANGUAGE'             'F'  0 2     '12'      5.0   4.9      4
     'SEX'                  'F'  0 2     '12'      5.0   4.9      4
     'NUMBER SERIES'        'F'  0 4     '1234'    5.0   4.9      4
     'FIGURE ANALOGIES'     'F'  0 4     '1234'    5.0   4.9      4
     'PATTERN COMPLETION'   'F'  0 4     '1234'    5.0   4.9      4
     'WORD PAIRS'           'F'  0 4     '1234'    5.0   4.9      4
     'VERBAL REASONING'     'F'  0 4     '1234'    5.0   4.9      4
     'WORD ANALOGIES'       'F'  0 4     '1234'    5.0   4.9      4
      0  3  0  1  1  30 5000                                     5
```

(ii) *Input format and observed data (the dependent variable is read in last)*

From (ii) above, note that the categories are designated by the numbers 1, 2, 3, 4 and that the number 0 is not permissible. Missing values should also be eliminated from the data set.

The order of the input contained in (i) above consists of the following five subsections:

Section 1—Description of dependent variable and number of categories of this variable. When using free format place description and number of categories between " " marks.

Section 2—Alphanumerical symbols for indicating levels (categories of the dependent variable). Note: Only one symbol per level.

Section 3—Number of predictors

Section 4—A, B, C, D, E, F, G

A = name of predictor

B = type of predictor: M = monotone,

F = free

C = Inclusion in analysis: 0 = included

1 = not permitted to subdivide

D = number of categories (levels)

E = symbols for the categories (alphanumerical)—one symbol per level

F = Level of significance for merging into groups

G = Level of significance for subdivision of groups

Section 5—Option number 1–7

Option 1 = 0   Print tables as row %

= 1   Print tables as %

= 2   Print tables as numbers

Option 2 = 0   Print all tables

= 1   Suppress "presubdivision" tables

= 2   Suppress "postsubdivision" tables

= 3   Print no tables

Option 3 = 0   Minimum number per subgroup = 50

> 20   Minimum number per subgroup

Option 4 = 0   5% level of significance for subdivision

= k   k% level of significance

Option 5 = 0   5% Bonferroni level of significance for subdivision

= k   k%-Bonferroni level of significance

Option 6 = 0   Maximum number of groups in dendrogram = 10

= m   Maximum number of groups = m

Option 7 = 0   Dimension of temporary vector = 500

= l   Dimension of temporary vector = l

A section of the CHAID output from which the dendrogram in Figure 8.3 was compiled, will now be shown.

Output:

```
CHAID   G.V.KASS.    V2

STUDY FIELD HAS  4 CATEGORIES CALLED:
1
2
3
4


THERE ARE  8 PREDICTORS AS FOLLOWS

TYPE      NO. OF CATEGORIES     CATEGORY SYMBOLS    USE?   MERGE%   SPLIT%    NAME
FREE              2                  12             MAY     5.00     4.90     LANGUAGE
FREE              2                  12             MAY     5.00     4.90     SEX
FREE              4                 1234            MAY     5.00     4.90     NUMBER SERIES
FREE              4                 1234            MAY     5.00     4.90     FIGURE ANALOGIES
FREE              4                 1234            MAY     5.00     4.90     PATTERN COMPLETION
FREE              4                 1234            MAY     5.00     4.90     WORD PAIRS
FREE              4                 1234            MAY     5.00     4.90     VERBAL REASONING
FREE              4                 1234            MAY     5.00     4.90     WORD ANALOGIES


OPTION   1   IS        0
OPTION   2   IS        3
OPTION   3   IS        0
OPTION   4   IS        1
OPTION   5   IS        1
OPTION   6   IS       30
OPTION   7   IS     5000
OPTION   8   IS        0
OPTION   9   IS        0
OPTION  10   IS        0


OPTIONS IN EFFECT:
TABLES PRINTED WITH ROW PERCENTAGES
DETAILED OUTPUT WILL BE SUPPRESSED
  TABLES NOT PRINTED
A GROUP WITH LESS THAN  50  CASES WILL NOT BE ANALYSED
A GROUP WILL NOT BE SPLIT IF IT IS NOT SIGNIFICANT AT THE  1.0000% LEVEL
A GROUP WILL NOT BE SPLIT IF IT IS NOT BONFERRONI SOGNIFICANT AT THE  1.0000% LEVEL
THE PROGRAM WILL TERMINATE WHEN 30 OR MORE GROUPS HAVE BEEN FORMED OR ANALYSED
INTERNAL WORK AREA <ICA> HAS LENGTH 5000
CHI-SQUARE IS STANDARD GOODNESS-OF-FIT STATISTIC


SUMMARY OF RESULTS
ANALYSIS OF TOTAL GROUP.

NO.      NAME                                                   SIGNIF (%)    BONF-SIG (%)  NUMBER OF GROUPS.
 1 LANGUAGE . . . . . . . . . . . . . . . . . . . . . . . .     4.0708990     4.0708990     2    1  2
 2 SEX . . . . . . . . . . . . . . . . . . . . . . . . . . .    3.7098E-48    3.7098E-48    2    1  2
 3 NUMBER SERIES . . . . . . . . . . . . . . . . . . . . . .    8.4519E-11    5.9164E-10    2    12  34
 4 FIGURE ANALOGIES . . . . . . . . . . . . . . . . . . . .    5.7252E-15    5.7252E-15    4    1  2  3  4
 5 PATTERN COMPLETION . . . . . . . . . . . . . . . . . . .    0.0001156     0.0006938     3    1  23  4
 6 WORD PAIRS . . . . . . . . . . . . . . . . . . . . . . .    5.3400E-24    5.3400E-24    4    1  2  3  4
 7 VERBAL REASONING . . . . . . . . . . . . . . . . . . . .    1.5399E-19    1.5399C-10    4    1  2  3  4
 8 WORD ANALOGIES . . . . . . . . . . . . . . . . . . . . .    4.0744E-16    4.0744E-16    4    1  2  3  4

CHARACTERISTICS OF THE BEST PREDICTOR
 2 SEX . . . . . . . . . . . . . . . . . . . . . . . . . . .    3.7098E-48    3.7098E-48    2    1  2

**********************************************************************************************************
**********
PREDICTOR   2:   SEX * ROW PERCENTAGES *
                            1        2     TOTAL
                   1     14.4      7.8     11.6
                   2     17.7     17.1     17.4
                   3      2.3     18.7      9.2
                   4     65.7     56.3     61.8
        TOTALS (100%)    1595     1152     2747
                            TABLE IS SIGNIFICANT AT  3.7098E-48% LEVEL
```

```
SUMMARY OF RESULTS
ANALYSIS OF SUBGROUP SEX (1)

NO.     NAME                                        SIGNIF (%)   BONF-SIG (%)  NUMBER OF GROUPS.
 1 LANGUAGE . . . . . . . . . . . . . . . . . . .  100.0000     100.0000      1   12
 2 SEX . . . . . . . . . . . . . . . . . . . . .   100.0000     100.0000      1   1
 3 NUMBER SERIES . . . . . . . . . . . . . . . .   2.3603E-07   1.4162E-06    3   1   2  34
 4 FIGURE ANALOGIES . . . . . . . . . . . . . .    4.7562E-09   2.8537E-08    3   1  23   4
 5 PATTERN COMPLETION . . . . . . . . . . . . .    3.7469E-05   0.0002248     3   12   3   4
 6 WORD PAIRS . . . . . . . . . . . . . . . . .    1.8226E-15   1.8226E-15    4   1   2   3   4
 7 VERBAL REASONING . . . . . . . . . . . . . .    1.6715E-15   1.0029E-14    3   12   3   4
 8 WORD ANALOGIES . . . . . . . . . . . . . . .    9.7837E-09   5.8702E-08    3   1  23   4

CHARACTERISTICS OF THE BEST PREDICTOR
 6 WORD PAIRS . . . . . . . . . . . . . . . . .    1.8226E-15   1.8226E-15    4   1   2   3   4

**********************************************************************************************
**********
PREDICTOR   6:  WORD PAIRS * ROW PERCENTAGES *
                      1      2      3      4     TOTAL
             1      5.8   13.6   17.0   26.0     14.4
             2     14.3   14.4   20.0   24.3     17.7
             3      2.4    1.8    2.4    2.3      2.3
             4     77.5   70.2   60.6   47.3     65.7
   TOTALS (100%)   502    382    411    300      1595
                                  TABLE IS SIGNIFICANT AT   1.8226E-15% LEVEL
SUMMARY OF RESULTS
ANALYSIS OF SUBGROUP SEX (2)

NO.     NAME                                        SIGNIF (%)   BONF-SIG (%)  NUMBER OF GROUPS.
 1 LANGUAGE . . . . . . . . . . . . . . . . . . .  0.0279542    0.0279542     2   1   2
 2 SEX . . . . . . . . . . . . . . . . . . . . .   100.0000     100.0000      1   2
 3 NUMBER SERIES . . . . . . . . . . . . . . . .   0.0130424    0.0782545     3   1  23   4
 4 FIGURE ANALOGIES . . . . . . . . . . . . . .    0.0014620    0.0102343     2   12  34
 5 PATTERN COMPLETION . . . . . . . . . . . . .    0.3844855    2.6913958     2   1  234
 6 WORD PAIRS . . . . . . . . . . . . . . . . .    1.0527E-08   6.3162E-08    3   12   3   4
 7 VERBAL REASONING . . . . . . . . . . . . . .    0.0006385    0.0038308     3   1   2  34
 8 WORD ANALOGIES . . . . . . . . . . . . . . .    3.7502E-06   2.2501E-05    3   1  23   4

CHARACTERISTICS OF THE BEST PREDICTOR
 6 WORD PAIRS . . . . . . . . . . . . . . . . .    1.0527E-08   6.3162E-08    3   12   3   4

**********************************************************************************************
**********
PREDICTOR   6:  WORD PAIRS * ROW PERCENTAGES *
                    1,2      3      4     TOTAL
             1      4.4   10.2   12.3     7.8
             2     12.4   17.2   26.8    17.1
             3     23.0   14.9   14.1    18.7
             4     60.2   57.8   46.7    56.3
   TOTALS (100%)   573    303    276      1152
                            TABLE IS SIGNIFICANT AT   1.0527E-08% LEVEL
SUMMARY OF RESULTS
ANALYSIS OF SUBGROUP SEX (1)  WORD PAIRS (1)

NO.     NAME                                        SIGNIF (%)   BONF-SIG (%)  NUMBER OF GROUPS.
 1 LANGUAGE . . . . . . . . . . . . . . . . . . .  100.0000     100.0000      1   12
 2 SEX . . . . . . . . . . . . . . . . . . . . .   100.0000     100.0000      1   1
 3 NUMBER SERIES . . . . . . . . . . . . . . . .   3.8973770    27.281601     2   134   2
 4 FIGURE ANALOGIES . . . . . . . . . . . . . .    100.0000     100.0000      1   1234
 5 PATTERN COMPLETION . . . . . . . . . . . . .    0.0016257    0.0097543     3   12   3   4
 6 WORD PAIRS . . . . . . . . . . . . . . . . .    100.0000     100.0000      1   1
 7 VERBAL REASONING . . . . . . . . . . . . . .    100.0000     100.0000      1   1234
 8 WORD ANALOGIES . . . . . . . . . . . . . . .    100.0000     100.0000      1   1423

CHARACTERISTICS OF THE BEST PREDICTOR
 5 PATTERN COMPLETION . . . . . . . . . . . . .    0.0016257    0.0097543     3   12  '3   4

**********************************************************************************************
**********
PREDICTOR   5:  PATTERN COMPLETION * ROW PERCENTAGES *
                    1,2      3      4     TOTAL
             1      4.1    3.3   17.4     5.8
             2     18.2    5.4    7.2    14.3
             3      2.1    3.3    2.9     2.4
             4     75.7   88.0   72.5    77.5
   TOTALS (100%)   341     92     69      502
                            TABLE IS SIGNIFICANT AT   0.0016257% LEVEL
SUMMARY OF RESULTS
ANALYSIS OF SUBGROUP SEX (1)  WORD PAIRS (2)

NO.     NAME                                        SIGNIF (%)   BONF-SIG (%)  NUMBER OF GROUPS.
 1 LANGUAGE . . . . . . . . . . . . . . . . . . .  0.1785790    0.1785790     2   1   2
 2 SEX . . . . . . . . . . . . . . . . . . . . .   100.0000     100.0000      1   1
 3 NUMBER SERIES . . . . . . . . . . . . . . . .   100.0000     100.0000      1   1243
 4 FIGURE ANALOGIES . . . . . . . . . . . . . .    100.0000     100.0000      1   1234
 5 PATTERN COMPLETION . . . . . . . . . . . . .    100.0000     100.0000      1   1234
 6 WORD PAIRS . . . . . . . . . . . . . . . . .    100.0000     100.0000      1   2
 7 VERBAL REASONING . . . . . . . . . . . . . .    1.9363184    13.5541954    2   12  34
 8 WORD ANALOGIES . . . . . . . . . . . . . . .    100.0000     100.0000      1   1342

CHARACTERISTICS OF THE BEST PREDICTOR
 1 LANGUAGE . . . . . . . . . . . . . . . . . .    0.1785790    0.1785790     2   1   2

**********************************************************************************************
**********
PREDICTOR   1:  LANGUAGE * ROW PERCENTAGES *
                      1      2     TOTAL
             1      8.8   19.2     13.6
             2     11.2   18.1     14.4
             3      2.4    1.1      1.8
             4     77.6   61.6     70.2
   TOTALS (100%)   205    177      382
                            TABLE IS SIGNIFICANT AT   0.1785790% LEVEL
SUMMARY OF RESULTS
ANALYSIS OF SUBGROUP SEX (1)  WORD PAIRS (3)

NO.     NAME                                        SIGNIF (%)   BONF-SIG (%)  NUMBER OF GROUPS.
 1 LANGUAGE . . . . . . . . . . . . . . . . . . .  100.0000     100.0000      1   12
 2 SEX . . . . . . . . . . . . . . . . . . . . .   100.0000     100.0000      1   1
 3 NUMBER SERIES . . . . . . . . . . . . . . . .   0.2048329    1.4338284     2   12  34
 4 FIGURE ANALOGIES . . . . . . . . . . . . . .    4.2813339    29.9692841    2   12  43
 5 PATTERN COMPLETION . . . . . . . . . . . . .    100.0000     100.0000      1   1234
 6 WORD PAIRS . . . . . . . . . . . . . . . . .    100.0000     100.0000      1   3
 7 VERBAL REASONING . . . . . . . . . . . . . .    0.1525747    1.0680218     2   12  34
 8 WORD ANALOGIES . . . . . . . . . . . . . . .    100.0000     100.0000      1   1234

CHARACTERISTICS OF THE BEST PREDICTOR
 7 VERBAL REASONING . . . . . . . . . . . . . .    0.1525747    1.0680218     2   12  34

THIS PREDICTOR IS NOT SIGNIFICANT.
```

```
SUMMARY OF RESULTS
ANALYSIS OF SUBGROUP SEX (1)  WORD PAIRS (4)

NO.      NAME                                      SIGNIF (%)   BONF-SIG (%) NUMBER OF GROUPS.
 1 LANGUAGE . . . . . . . . . . . . . . . . . .    100.0000     100.0000    1   12
 2 SEX . . . . . . . . . . . . . . . . . . . . .   100.0000     100.0000    1   1
 3 NUMBER SERIES . . . . . . . . . . . . . . . .     0.5508225    3.8557539  2   123  4
 4 FIGURE ANALOGIES . . . . . . . . . . . . . .      0.0878501    0.5270997  3   1  23 4
 5 PATTERN COMPLETION . . . . . . . . . . . . .      1.6845961   11.7921486  2   13  24
 6 WORD PAIRS . . . . . . . . . . . . . . . . .    100.0000     100.0000    1   4
 7 VERBAL REASONING . . . . . . . . . . . . . .      0.1456944    1.0198593  2   123  4
 8 WORD ANALOGIES . . . . . . . . . . . . . . .      0.5438933    3.8072472  2   123  4

CHARACTERISTICS OF THE BEST PREDICTOR
 4 FIGURE ANALOGIES . . . . . . . . . . . . . .      0.0878501    0.5270997  3   1  23 4
```

```
**********
PREDICTOR  4:  FIGURE ANALOGIES * ROW PERCENTAGES *
                           1    2,3    4    TOTAL
                  1      2.5   23.8  40.2    26.0
                  2     37.5   23.2  20.7    24.3
                  3      2.5    2.4   2.2     2.3
                  4     57.5   50.6  37.0    47.3
       TOTALS (100%)     40    168    92     300
                    TABLE IS SIGNIFICANT AT   0.0878501% LEVEL
```

```
SUMMARY OF RESULTS
ANALYSIS OF SUBGROUP SEX (2)  WORD PAIRS (1,2)

NO.      NAME                                      SIGNIF (%)   BONF-SIG (%) NUMBER OF GROUPS.
 1 LANGUAGE . . . . . . . . . . . . . . . . . .    100.0000     100.0000    1   12
 2 SEX . . . . . . . . . . . . . . . . . . . . .   100.0000     100.0000    1   2
 3 NUMBER SERIES . . . . . . . . . . . . . . . .     3.8241720   26.7691650  2   13  24
 4 FIGURE ANALOGIES . . . . . . . . . . . . . .      3.2034292   22.4239655  2   13  24
 5 PATTERN COMPLETION . . . . . . . . . . . . .    100.0000     100.0000    1   1243
 6 WORD PAIRS . . . . . . . . . . . . . . . . .    100.0000     100.0000    1   12
 7 VERBAL REASONING . . . . . . . . . . . . . .      0.0120502    0.0843515  2   12  34
 8 WORD ANALOGIES . . . . . . . . . . . . . . .      1.2320967    8.6246634  2   13  24

CHARACTERISTICS OF THE BEST PREDICTOR
 7 VERBAL REASONING . . . . . . . . . . . . . .      0.0120502    0.0843515  2   12  34
```

```
**********
PREDICTOR  7:  VERBAL REASONING * ROW PERCENTAGES *
                          1,2    3,4   TOTAL
                  1      1.8   10.1     4.4
                  2     12.9   11.2    12.4
                  3     24.4   20.1    23.0
                  4     60.9   58.7    60.2
       TOTALS (100%)    394    179     573
                    TABLE IS SIGNIFICANT AT   0.0120502% LEVEL
```

```
SUMMARY OF RESULTS
ANALYSIS OF SUBGROUP SEX (2)  WORD PAIRS (3)

NO.      NAME                                      SIGNIF (%)   BONF-SIG (%) NUMBER OF GROUPS.
 1 LANGUAGE . . . . . . . . . . . . . . . . . .    100.0000     100.0000    1   12
 2 SEX . . . . . . . . . . . . . . . . . . . . .   100.0000     100.0000    1   2
 3 NUMBER SERIES . . . . . . . . . . . . . . . .   100.0000     100.0000    1   1234
 4 FIGURE ANALOGIES . . . . . . . . . . . . . .    100.0000     100.0000    1   1234
 5 PATTERN COMPLETION . . . . . . . . . . . . .    100.0000     100.0000    1   1423
 6 WORD PAIRS . . . . . . . . . . . . . . . . .    100.0000     100.0000    1   3
 7 VERBAL REASONING . . . . . . . . . . . . . .    100.0000     100.0000    1   1423
 8 WORD ANALOGIES . . . . . . . . . . . . . . .      1.5467043   10.8269053  2   1  234

CHARACTERISTICS OF THE BEST PREDICTOR
 8 WORD ANALOGIES . . . . . . . . . . . . . . .      1.5467043   10.8269053  2   1  234

THIS PREDICTOR IS NOT SIGNIFICANT.
```

```
SUMMARY OF RESULTS
ANALYSIS OF SUBGROUP SEX (2)  WORD PAIRS (4)

NO.      NAME                                      SIGNIF (%)   BONF-SIG (%) NUMBER OF GROUPS.
 1 LANGUAGE . . . . . . . . . . . . . . . . . .    100.0000     100.0000    1   12
 2 SEX . . . . . . . . . . . . . . . . . . . . .   100.0000     100.0000    1   2
 3 NUMBER SERIES . . . . . . . . . . . . . . . .     4.4698992   31.2892456  2   1  234
 4 FIGURE ANALOGIES . . . . . . . . . . . . . .    100.0000     100.0000    1   1234
 5 PATTERN COMPLETION . . . . . . . . . . . . .    100.0000     100.0000    1   1324
 6 WORD PAIRS . . . . . . . . . . . . . . . . .    100.0000     100.0000    1   4
 7 VERBAL REASONING . . . . . . . . . . . . . .    100.0000     100.0000    1   1234
 8 WORD ANALOGIES . . . . . . . . . . . . . . .    100.0000     100.0000    1   1324

CHARACTERISTICS OF THE BEST PREDICTOR
 3 NUMBER SERIES . . . . . . . . . . . . . . . .     4.4698992   31.2892456  2   1  234

THIS PREDICTOR IS NOT SIGNIFICANT.
```

```
SUMMARY OF RESULTS
ANALYSIS OF SUBGROUP SEX (1)  WORD PAIRS (1)  PATTERN COMPLETION (1,2)

NO.      NAME                                      SIGNIF (%)   BONF-SIG (%) NUMBER OF GROUPS.
 1 LANGUAGE . . . . . . . . . . . . . . . . . .    100.0000     100.0000    1   12
 2 SEX . . . . . . . . . . . . . . . . . . . . .   100.0000     100.0000    1   1
 3 NUMBER SERIES . . . . . . . . . . . . . . . .     0.1107638    0.7753453  2   1  234
 4 FIGURE ANALOGIES . . . . . . . . . . . . . .    100.0000     100.0000    1   1432
 5 PATTERN COMPLETION . . . . . . . . . . . . .    100.0000     100.0000    1   12
 6 WORD PAIRS . . . . . . . . . . . . . . . . .    100.0000     100.0000    1   1
 7 VERBAL REASONING . . . . . . . . . . . . . .      1.3992119    9.7944746  2   1  234
 8 WORD ANALOGIES . . . . . . . . . . . . . . .      2.5934191   18.1538849  2   12  34

CHARACTERISTICS OF THE BEST PREDICTOR
 3 NUMBER SERIES . . . . . . . . . . . . . . . .     0.1107638    0.7753453  2   1  234
```

```
**********
PREDICTOR  3:  NUMBER SERIES * ROW PERCENTAGES *
                          1    2,3,4   TOTAL
                  1      2.5    5.6     4.1
                  2     10.6   25.0    10.2
                  3      3.1    1.1     2.1
                  4     83.9   68.3    75.7
       TOTALS (100%)    161    180     341
                    TABLE IS SIGNIFICANT AT   0.1107638% LEVEL
```

```
SUMMARY OF RESULTS
ANALYSIS OF SUBGROUP SEX (1)  WORD PAIRS (1)  PATTERN COMPLETION (3)

NO.      NAME                                      SIGNIF (%)   BONF-SIG (%) NUMBER OF GROUPS.
 1 LANGUAGE . . . . . . . . . . . . . . . . . .    100.0000     100.0000    1   12
 2 SEX . . . . . . . . . . . . . . . . . . . . .   100.0000     100.0000    1   1
 3 NUMBER SERIES . . . . . . . . . . . . . . . .    100.0000     100.0000    1   1342
 4 FIGURE ANALOGIES . . . . . . . . . . . . . .    100.0000     100.0000    1   1423
 5 PATTERN COMPLETION . . . . . . . . . . . . .    100.0000     100.0000    1   3
 6 WORD PAIRS . . . . . . . . . . . . . . . . .    100.0000     100.0000    1   1
 7 VERBAL REASONING . . . . . . . . . . . . . .    100.0000     100.0000    1   1423
 8 WORD ANALOGIES . . . . . . . . . . . . . . .    100.0000     100.0000    1   1234

NO PREDICTION IS POSSIBLE.
```

```
SUMMARY OF RESULTS
ANALYSIS OF SUBGROUP SEX (1)   WORD PAIRS (1)   PATTERN COMPLETION (4)

NO.      NAME                                        SIGNIF (%)   BONF-SIG (%)  NUMBER OF GROUPS.
 1 LANGUAGE . . . . . . . . . . . . . . . . . . . .  100.0000     100.0000    1    12
 2 SEX . . . . . . . . . . . . . . . . . . . . . . . 100.0000     100.0000    1    1
 3 NUMBER SERIES . . . . . . . . . . . . . . . . . .   1.6924343    11.8470097 2    143   2
 4 FIGURE ANALOGIES . . . . . . . . . . . . . . . .  100.0000     100.0000    1    1243
 5 PATTERN COMPLETION . . . . . . . . . . . . . . .  100.0000     100.0000    1    1234
 6 WORD PAIRS . . . . . . . . . . . . . . . . . . .  100.0000     100.0000    1    1
 7 VERBAL REASONING . . . . . . . . . . . . . . . .  100.0000     100.0000    1    1432
 8 WORD ANALOGIES . . . . . . . . . . . . . . . . .  100.0000     100.0000    1    1234

CHARACTERISTICS OF THE BEST PREDICTOR
 3 NUMBER SERIES . . . . . . . . . . . . . . . . . .   1.6924343    11.8470097 2    143   2

THIS PREDICTOR IS NOT SIGNIFICANT.
SUMMARY OF RESULTS
ANALYSIS OF SUBGROUP SEX (1)   WORD PAIRS (2)   LANGUAGE (1)

NO.      NAME                                        SIGNIF (%)   BONF-SIG (%)  NUMBER OF GROUPS.
 1 LANGUAGE . . . . . . . . . . . . . . . . . . . .  100.0000     100.0000    1    1
 2 SEX . . . . . . . . . . . . . . . . . . . . . . . 100.0000     100.0000    1    1
 3 NUMBER SERIES . . . . . . . . . . . . . . . . . .  100.0000     100.0000    1    1423
 4 FIGURE ANALOGIES . . . . . . . . . . . . . . . .  100.0000     100.0000    1    1423
 5 PATTERN COMPLETION . . . . . . . . . . . . . . .  100.0000     100.0000    1    1234
 6 WORD PAIRS . . . . . . . . . . . . . . . . . . .  100.0000     100.0000    1    2
 7 VERBAL REASONING . . . . . . . . . . . . . . . .  100.0000     100.0000    1    1243
 8 WORD ANALOGIES . . . . . . . . . . . . . . . . .  100.0000     100.0000    1    1342

NO PREDICTION IS POSSIBLE.
SUMMARY OF RESULTS
ANALYSIS OF SUBGROUP SEX (1)   WORD PAIRS (2)   LANGUAGE (2)

NO.      NAME                                        SIGNIF (%)   BONF-SIG (%)  NUMBER OF GROUPS.
 1 LANGUAGE . . . . . . . . . . . . . . . . . . . .  100.0000     100.0000    1    2
 2 SEX . . . . . . . . . . . . . . . . . . . . . . . 100.0000     100.0000    1    1
 3 NUMBER SERIES . . . . . . . . . . . . . . . . . .  100.0000     100.0000    1    1324
 4 FIGURE ANALOGIES . . . . . . . . . . . . . . . .  100.0000     100.0000    1    1234
 5 PATTERN COMPLETION . . . . . . . . . . . . . . .  100.0000     100.0000    1    1234
 6 WORD PAIRS . . . . . . . . . . . . . . . . . . .  100.0000     100.0000    1    2
 7 VERBAL REASONING . . . . . . . . . . . . . . . .    3.9604778    27.7232971 2    1   234
 8 WORD ANALOGIES . . . . . . . . . . . . . . . . .  100.0000     100.0000    1    1324

CHARACTERISTICS OF THE BEST PREDICTOR
 7 VERBAL REASONING . . . . . . . . . . . . . . . .    3.9604778    27.7232971 2    1   234

THIS PREDICTOR IS NOT SIGNIFICANT.
SUMMARY OF RESULTS
ANALYSIS OF SUBGROUP SEX (1)   WORD PAIRS (4)   FIGURE ANALOGIES (2,3)

NO.      NAME                                        SIGNIF (%)   BONF-SIG (%)  NUMBER OF GROUPS.
 1 LANGUAGE . . . . . . . . . . . . . . . . . . . .  100.0000     100.0000    1    12
 2 SEX . . . . . . . . . . . . . . . . . . . . . . . 100.0000     100.0000    1    1
 3 NUMBER SERIES . . . . . . . . . . . . . . . . . .   0.5938177     4.1567163 2    14    23
 4 FIGURE ANALOGIES . . . . . . . . . . . . . . . .  100.0000     100.0000    1    23
 5 PATTERN COMPLETION . . . . . . . . . . . . . . .  100.0000     100.0000    1    1342
 6 WORD PAIRS . . . . . . . . . . . . . . . . . . .  100.0000     100.0000    1    4
 7 VERBAL REASONING . . . . . . . . . . . . . . . .    3.1806192    22.2642975 2    14    23
 8 WORD ANALOGIES . . . . . . . . . . . . . . . . .   0.3080304     2.1562090 2    123   4

CHARACTERISTICS OF THE BEST PREDICTOR
 8 WORD ANALOGIES . . . . . . . . . . . . . . . . .   0.3080304     2.1562090 2    123   4

THIS PREDICTOR IS NOT SIGNIFICANT.
SUMMARY OF RESULTS
ANALYSIS OF SUBGROUP SEX (1)   WORD PAIRS (4)   FIGURE ANALOGIES (4)

NO.      NAME                                        SIGNIF (%)   BONF-SIG (%)  NUMBER OF GROUPS.
 1 LANGUAGE . . . . . . . . . . . . . . . . . . . .  100.0000     100.0000    1    12
 2 SEX . . . . . . . . . . . . . . . . . . . . . . . 100.0000     100.0000    1    1
 3 NUMBER SERIES . . . . . . . . . . . . . . . . . .  100.0000     100.0000    1    1234
 4 FIGURE ANALOGIES . . . . . . . . . . . . . . . .  100.0000     100.0000    1    4
 5 PATTERN COMPLETION . . . . . . . . . . . . . . .    3.2277374    22.5941467 2    14    23
 6 WORD PAIRS . . . . . . . . . . . . . . . . . . .  100.0000     100.0000    1    4
 7 VERBAL REASONING . . . . . . . . . . . . . . . .  100.0000     100.0000    1    1234
 8 WORD ANALOGIES . . . . . . . . . . . . . . . . .  100.0000     100.0000    1    1234

CHARACTERISTICS OF THE BEST PREDICTOR
 5 PATTERN COMPLETION . . . . . . . . . . . . . . .    3.2277374    22.5941467 2    14    23

THIS PREDICTOR IS NOT SIGNIFICANT.
```

From the dendrogram it appears that sex as predictor explains most of the variation in proposed study field (Y). At the second level of partitioning it was found that word pairs (X_{16}) was the most important predictor. Note that for sex = 2, i.e. for schoolgirls, categories (1, 2) were grouped together to form a subgroup. At the third level of partitioning pattern completion (X_{15}), language group, figure analogies (X_{14}) and verbal reasoning (X_{17}) evidently all contributed to the variation in Y. The fact that each of these variables explain most of the variation in respect of a specific subpopulation, points to the presence of interactions between these predictors and the preceding predictors.

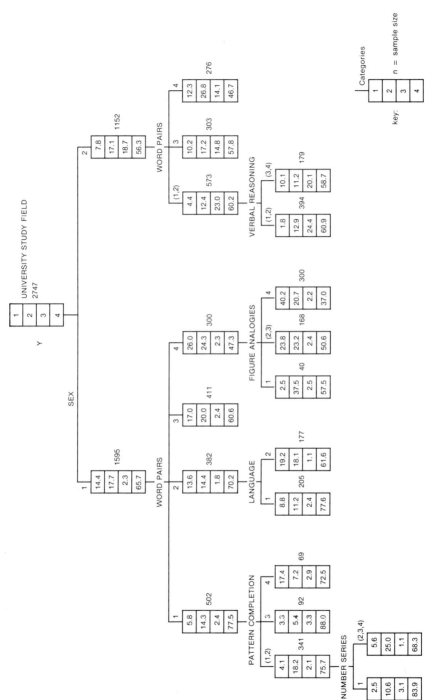

Figure 8.3. Dendrogram for proposed university study field.

4. XAID—An Exploratory Technique for Analyzing a Quantitative Dependent Variable with Categorical Predictors

As in the case of CHAID, XAID is preferably used for analyzing data sets consisting of a large number of observations. However in the case of XAID the dependent variable is quantitative in nature. The predictors are of a categorical nature and the values 1, 2, 3, ... are assigned to the various categories.

Three types of predictors are permissible, namely free, monotone and free in Category One. A brief description of each type of predictor will now be given:

Free: The categories of this predictor may be grouped in any order (nominal scale).

Monotone: The categories of this variable form an ascending or a descending order (ordinal scale).

Free in Category One: All categories, with the exception of Category One, are monotone. Category One may be placed in any position with regard to the remaining categories. Situations in which this type of predictor can be used meaningfully are as follows: (i) Where a specific predictor has many missing values, or (ii) where one of the categories of the predictor is indicated by "uncertain", whereas the remaining categories are ordinal. The missing values (Case (i)) or the category "uncertain" (Case (ii)) are placed in Category One.

As in the case of CHAID, the data set is divided into subgroups according to each predictor's ability to explain the variation in the dependent variable. Suppose a particular predictor consists of categories A, B, C, D, E. If this predictor divides the data set into two subgroups consisting of $\{A, B, E\}$ and $\{C, D\}$, this partitioning is then evaluated by means of a one-way analysis of variance with regard to the dependent variable Y and the two "treatments" $\{A, B, E\}$ and $\{C, D\}$.

The most significant predictor divides the data into subgroups at the first level of partitioning. The XAID procedure is then repeated in order to form further subgroups at the second, third, etc. levels of partitioning. The procedure ends when no further significant partitionings are possible.

Subgroups are formed as follows in respect of each predictor:

(i) Each category of a predictor initially forms a subgroup and t-tests are used for deciding which categories may be merged to form homogeneous subgroups. The process is continued until all subgroups consist of categories of which the mean Y values do not differ significantly.

(ii) Next in line is the partitioning phase. Each subgroup consisting of two categories or more is divided into two parts and then the t-values for

testing differences between means for significance are compared with
the previously specified partitioning (split) value. Groups are partitioned
according to the most significant t-value.

Steps (i) and (ii) are then repeated until no further merging or partitioning
occurs. The final grouping of the categories of the various predictors is used
for constructing the dendrogram. Note that the level of significance for partitioning should always be smaller than the corresponding merging value.
Typical values are 0.045 and 0.050, respectively.

5. Application of XAID Analysis

The full Ability Test Data set ($n = 2703$) as previousiy described in Chapter 1
was used for illustrating the application of XAID. As dependent variabie Y,
X_{20} was taken, namely, the Senior Aptitude Test (verbal). The five predictors
are language group (A, B), sex (male, female), number series (X_{13}), figure
analogies (X_{14}) and verbal reasoning (X_{17}). The predictors X_{13}, X_{14} and X_{17}
all refer to Educational Level III and were transformed into ordinal variables,
firstly by calculating the quartiles and then placing the outcome of each
predictor into four categories according to the method explained in Section 3.

Input: XAID—Particulars of variables and program options, followed by the
observed data

```
                                                              Section
 6  0  6        (6F3.0)            'SENIOR APTITUDE TEST'       1
 0.01  0                                                        2
 1  2 1-1      .0450      .0500    LANGUAGE                     3
 2  2 1-1      .0450      .0500    SEX                          3
 3  4 1 0      .0450      .0500    NUMBER SERIES                3
 4  4 1 0      .0450      .0500    FIGURE ANALOGIES             3
 5  4 1 0      .0450      .0500    VERBAL REASONING             3
   1  1  4  4  4 19                                             4
   1  1  1  3  3 15
   1  1  2  3  4 16
   1  1  3  1  3 10
   1  1  1  3  2  7
   1  1  1  4  2 18
                      .
                      .
```

The input consists of the following four suosections described in Heymann
(1981).

Section 1. N, IW, IDEP, FORM, NAMDEP

Columns 1–2:   N = Total number of variables including the dependent
variable and a weighting variable, if cases are weighted.

Columns 3–4:   IW = The number of the weighting variable. For equal weighting substitute IW = 0.

Columns 5–6:   IDEP = The number of the dependent variable.

Columns 7–36:   FORM = The FORTRAN input format. The format is placed between brackets.

Columns 37–66: NAMDEP = The name of the dependent variable.

Section 2. SIG and IOPT (in free format)

SIG = The level of significance (not as a percentage) for evaluating the analysis.

IOPT = 0: Suppress output on FORTRAN writing unit 10. This file contains full information on the split/merge process in respect of each predictor.

IOPT = 1: Print all tables.

Section 3. IPRED, NCAT, MAY, NTYP, SPLIT, SMERG, NAME

The following information is required for each predictor:

Columns 1–2:   IPRED = The number of the variable (i.e., the column number in the data set corresponding to the particular variable).

Columns 3–4:   NCAT = The number of categories.

Columns 5–6:   MAY = 1 If the predictor may be used for partitioning.
                    = 0 Otherwise

Columns 7–8:   NTYP = −1: Free
                    = 0: Monotone
                    = 1: Free in Category One

Columns 9–18:   SPLIT = Level of significance of partitioning

Columns 19–28: SMERG = Level of significance of merging

Columns 29–58: NAME = Name of variable (a maximum of 30 characters)

Section 4. Observed data

Limitations of Program

(i) A maximum of 62 variables (one dependent variable, one weighting variable and 60 predictors) may be used.

(ii) The categories of the predictors may not be designated with the number 0, but only with 1, 2, 3, ... up to a maximum of 20, starting with 1.

(iii) The level of significance of partitioning must be smaller than the level of significance for the merging of groups.

A section of the computer output which is required for compiling the dendrogram (Figure 8.4) is as follows:

Output: Senior Aptitude Test with language, sex, numbers series, figure analogies and verbal reasoning as predictors

```
OVERALL SIGNIFICANCE FOR TESTS OF GROUPING  0.01000
NUMBER OF CASES        2703
DEPENDENT VARIABLE NO.    6 NAME=SENIOR APTITUDE TEST'
WEIGHT VARIABLE NO.    0

PREDICTOR VARIABLES:
NO                       NAME        NO.OF CAT.S    SPLIT      MERGE      MAY?    TYPE
 1              LANGUAGE                    2       0.0450     0.0500     YES     FREE
 2              SEX                         2       0.0450     0.0500     YES     FREE
 3              NUMBER SERIES               4       0.0450     0.0500     YES     MONO
 4              FIGURE ANALOGIES            4       0.0450     0.0500     YES     MONO
 5              VERBAL REASONING            4       0.0450     0.0500     YES     MONO

     ANALYSIS OF GROUP NO.  1          PREVIOUS GROUP NO.  0
NO.        NAME              MC-SIG(%)    BON-F-SIG(%)      GROUPING
 1     LANGUAGE              0.727D-05     0.218D-04      1  /  2
 2     SEX                   0.526D-20     0.158D-19      1  /  2
 3     NUMBER SERIES         0.100D-75     0.100D-75      1  /  2  /  3  /  4
 4     FIGURE ANALOGIES      0.224D-49     0.224D-49      1  /  2  /  3  /  4
 5     VERBAL REASONING      0.100D-75     0.100D-75      1  /  2  /  3  /  4

     BEST PREDICTOR
 5     VERBAL REASONING      0.100D-75     0.100D-75      1  /  2  /  3  /  4

STATISTICS FOR GROUPING:
GROUP NO.    MEAN  SAMPLE VARIANCE  NO.OF CASES    WEIGHT   WEIGHTED MEAN   VARIANCE(WEIGHTED MEAN)

   2      15.2450    11.1467          702       702.0000      15.2450          0.0159

   3      17.1874    10.2014          859       859.0000      17.1874          0.0119

   4      18.5381     9.3767          565       565.0000      18.5381          0.0166

   5      20.6291     9.1018          577       577.0000      20.6291          0.0158

     ANALYSIS OF VARIANCE:
                 SUM OF SQUARES      MEAN SQUARE      DEGREES OF FREEDOM
     GROUPING       9803.9257        3267.9752               3
     ERROR         27097.7456          10.0399            2699

     F-VALUE     =     325.50
     SIGNIFICANCE= 0.100D-75
     BON-F-SIG.  = 0.100D-75
     MC-SIG.     = 0.100D-75
     CONSERV.SIG.= 0.100D-75
     GROUPING IS SIGNIFICANT AT THE   0.100D-75%- LEVEL(CONSERVATIVE)

     ANALYSIS OF GROUP NO.  2          PREVIOUS GROUP NO.  1
NO.        NAME              MC-SIG(%)    BON-F-SIG(%)      GROUPING
 1     LANGUAGE              0.358D-03     0.107D-02      1  /  2
 2     SEX                   0.174D-03     0.521D-03      1  /  2
 3     NUMBER SERIES         0.924D-06     0.526D-06      1  /  2  /  3  4
 4     FIGURE ANALOGIES      0.368D+01     0.106D+01      1  /  2  3  4
 5     VERBAL REASONING      0.100D+03     0.100D+03      1

     BEST PREDICTOR
 3     NUMBER SERIES         0.924D-06     0.526D-06      1  /  2  /  3  4

STATISTICS FOR GROUPING:
GROUP NO.    MEAN  SAMPLE VARIANCE  NO.OF CASES    WEIGHT   WEIGHTED MEAN   VARIANCE(WEIGHTED MEAN)

   6      14.5573    11.2527          375       375.0000      14.5573          0.0300

   7      15.7397    10.4319          219       219.0000      15.7397          0.0476

   8      16.6296     8.3475          108       108.0000      16.6296          0.0773

     ANALYSIS OF VARIANCE:
                 SUM OF SQUARES      MEAN SQUARE      DEGREES OF FREEDOM
     GROUPING        437.9905         218.9953               2
     ERROR          7375.8673          10.5520             699

     F-VALUE     =      20.754
     SIGNIFICANCE= 0.175D-06
     BON-F-SIG.  = 0.526D-06
     MC-SIG.     = 0.924D-06
     CONSERV.SIG.= 0.526D-06
     GROUPING IS SIGNIFICANT AT THE   0.526D-06%- LEVEL(CONSERVATIVE)

     ANALYSIS OF GROUP NO.  3          PREVIOUS GROUP NO.  1
NO.        NAME              MC-SIG(%)    BON-F-SIG(%)      GROUPING
 1     LANGUAGE              0.374D+00     0.112D+01      1  /  2
 2     SEX                   0.610D-10     0.183D-09      1  /  2
 3     NUMBER SERIES         0.226D-02     0.255D-03      1  2  /  3  4
 4     FIGURE ANALOGIES      0.837D-01     0.134D-01      1  /  2  3  4
 5     VERBAL REASONING      0.100D+03     0.100D+03      2

     BEST PREDICTOR
 2     SEX                   0.610D-10     0.183D-09      1  /  2

STATISTICS FOR GROUPING:
GROUP NO.    MEAN  SAMPLE VARIANCE  NO.OF CASES    WEIGHT   WEIGHTED MEAN   VARIANCE(WEIGHTED MEAN)

   9      16.5368     9.5081          503       503.0000      16.5368          0.0189

  10      18.1067     9.7632          356       356.0000      18.1067          0.0274

     ANALYSIS OF VARIANCE:
                 SUM OF SQUARES      MEAN SQUARE      DEGREES OF FREEDOM
     GROUPING        513.8108         513.8108               1
     ERROR          8239.0137           9.6138             857

     F-VALUE     =      53.445
     SIGNIFICANCE= 0.610D-10
     BON-F-SIG.  = 0.183D-09
     MC-SIG.     = 0.610D-10
     CONSERV.SIG.= 0.610D-10
     GROUPING IS SIGNIFICANT AT THE   0.610D-10%- LEVEL(CONSERVATIVE)
```

```
      ANALYSIS OF GROUP NO.  4          PREVIOUS GROUP NO.  1
NO.        NAME                MC-SIG(%)   BON-F-SIG(%)      GROUPING
 1      LANGUAGE              0.104D+01    0.313D+01     1  /  2
 2      SEX                   0.520D-04    0.156D-03     1  /  2
 3      NUMBER SERIES         0.168D+00    0.412D+00     1  /  2  3  4
 4      FIGURE ANALOGIES      0.121D+01    0.281D+00     1  2  /  3  4
 5      VERBAL REASONING      0.100D+03    0.100D+03     3


   BEST PREDICTOR
 2      SEX                   0.520D-04    0.156D-03     1  /  2

STATISTICS FOR GROUPING:
GROUP NO.    MEAN  SAMPLE VARIANCE  NO.OF CASES    WEIGHT    WEIGHTED MEAN   VARIANCE(WEIGHTED MEAN)

   11   17.9511     8.7590          307       307.0000       17.9511           0.0285

   12   19.2364     9.2474          258       258.0000       19.2364           0.0358


   ANALYSIS OF VARIANCE:
                 SUM OF SQUARES       MEAN SQUARE      DEGREES OF FREEDOM
   GROUPING         231.5873         231.5873                 1
   ERROR           5056.8449           8.9820               563

   F-VALUE     =     25.784
   SIGNIFICANCE=  0.520D-04
   BON-F-SIG.  =  0.156D-03
   MC-SIG.     =  0.520D-04
   CONSERV.SIG.=  0.520D-04
   GROUPING IS SIGNIFICANT AT THE   0.520D-04%- LEVEL(CONSERVATIVE)


      ANALYSIS OF GROUP NO.  5          PREVIOUS GROUP NO.  1
NO.        NAME                MC-SIG(%)   BON-F-SIG(%)      GROUPING
 1      LANGUAGE              0.808D-02    0.242D-01     1  /  2
 2      SEX                   0.607D-05    0.182D-04     1  /  2
 3      NUMBER SERIES         0.153D-06    0.959D-08     1  2  3  /  4
 4      FIGURE ANALOGIES      0.291D+00    0.542D-01     1  2  /  3  4
 5      VERBAL REASONING      0.100D+03    0.100D+03     4


   BEST PREDICTOR
 3      NUMBER SERIES         0.153D-06    0.959D-08     1  2  3  /  4

STATISTICS FOR GROUPING:
GROUP NO.    MEAN  SAMPLE VARIANCE  NO.OF CASES    WEIGHT    WEIGHTED MEAN   VARIANCE(WEIGHTED MEAN)

   13   19.8557     8.9396          305       305.0000       19.8557           0.0293

   14   21.4963     7.8893          272       272.0000       21.4963           0.0290


   ANALYSIS OF VARIANCE:
                 SUM OF SQUARES       MEAN SQUARE      DEGREES OF FREEDOM
   GROUPING         386.9821         386.9821                 1
   ERROR           4855.6490           8.4446               575

   F-VALUE     =     45.826
   SIGNIFICANCE=  0.320D-08
   BON-F-SIG.  =  0.959D-08
   MC-SIG.     =  0.153D-06
   CONSERV.SIG.=  0.959D-08
   GROUPING IS SIGNIFICANT AT THE   0.959D-08%- LEVEL(CONSERVATIVE)


      ANALYSIS OF GROUP NO.  6          PREVIOUS GROUP NO.  2
NO.        NAME                MC-SIG(%)   BON-F-SIG(%)      GROUPING
 1      LANGUAGE              0.418D-02    0.125D-01     1  /  2
 2      SEX                   0.380D-02    0.114D-01     1  /  2
 3      NUMBER SERIES         0.100D+03    0.100D+03     1
 4      FIGURE ANALOGIES      0.100D+03    0.100D+03     1  2  3  4
 5      VERBAL REASONING      0.100D+03    0.100D+03     1


   BEST PREDICTOR
 2      SEX                   0.380D-02    0.114D-01     1  /  2

STATISTICS FOR GROUPING:
GROUP NO.    MEAN  SAMPLE VARIANCE  NO.OF CASES    WEIGHT    WEIGHTED MEAN   VARIANCE(WEIGHTED MEAN)

   15   14.0086    10.5321          234       234.0000       14.0086           0.0450

   16   15.4681    11.1936          141       141.0000       15.4681           0.0794


   ANALYSIS OF VARIANCE:
                 SUM OF SQUARES       MEAN SQUARE      DEGREES OF FREEDOM
   GROUPING         187.4281         187.4281                 1
   ERROR           4021.0899          10.7804               373

   F-VALUE     =     17.386
   SIGNIFICANCE=  0.380D-02
   BON-F-SIG.  =  0.114D-01
   MC-SIG.     =  0.380D-02
   CONSERV.SIG.=  0.380D-02
   GROUPING IS SIGNIFICANT AT THE   0.380D-02%- LEVEL(CONSERVATIVE)


      ANALYSIS OF GROUP NO.  7          PREVIOUS GROUP NO.  2
NO.        NAME                MC-SIG(%)   BON-F-SIG(%)      GROUPING
 1      LANGUAGE              0.100D+03    0.100D+03     1  2
 2      SEX                   0.965D-01    0.289D+00     1  /  2
 3      NUMBER SERIES         0.100D+03    0.100D+03     2
 4      FIGURE ANALOGIES      0.100D+03    0.100D+03     1  2  3  4
 5      VERBAL REASONING      0.100D+03    0.100D+03     1


   BEST PREDICTOR
 2      SEX                   0.965D-01    0.289D+00     1  /  2

STATISTICS FOR GROUPING:
GROUP NO.    MEAN  SAMPLE VARIANCE  NO.OF CASES    WEIGHT    WEIGHTED MEAN   VARIANCE(WEIGHTED MEAN)

   17   15.0609     9.7594          115       115.0000       15.0609           0.0849

   18   16.4904    10.1941          104       104.0000       16.4904           0.0980


   ANALYSIS OF VARIANCE:
                 SUM OF SQUARES       MEAN SQUARE      DEGREES OF FREEDOM
   GROUPING         111.6001         111.6001                 1
   ERROR           2162.5640           9.9657               217

   F-VALUE     =     11.198
   SIGNIFICANCE=  0.965D-01
   BON-F-SIG.  =  0.289D+00
   MC-SIG.     =  0.965D-01
   CONSERV.SIG.=  0.965D-01
   GROUPING IS SIGNIFICANT AT THE   0.965D-01%- LEVEL(CONSERVATIVE)
```

```
         ANALYSIS OF GROUP NO.  8          PREVIOUS GROUP NO.  2
NO.         NAME                 MC-SIG(%)   BON-F-SIG(%)      GROUPING
1           LANGUAGE             0.100D+03    0.100D+03        1   2
2           SEX                  0.100D+03    0.100D+03        2   1
3           NUMBER SERIES        0.100D+03    0.100D+03        3   4
4           FIGURE ANALOGIES     0.100D+03    0.100D+03        1   2   3   4
5           VERBAL REASONING     0.100D+03    0.100D+03        1


         BEST PREDICTOR
5           VERBAL REASONING     0.100D+03    0.100D+03        1


STATISTICS FOR GROUPING:
GROUP NO.    MEAN   SAMPLE VARIANCE   NO.OF CASES    WEIGHT    WEIGHTED MEAN   VARIANCE(WEIGHTED MEAN)

    19     16.6296     8.3475           108       108.0000      16.6296            0.0773


         ANALYSIS OF VARIANCE:
                    SUM OF SQUARES     MEAN SQUARE      DEGREES OF FREEDOM
         GROUPING         0.0             0.0                  0
         ERROR            0.0             0.0                  0

         F-VALUE    =      .0
         SIGNIFICANCE=  0.100D+03
         BON-F-SIG.  =  0.100D+03
         MC-SIG.     =  0.100D+03
         CONSERV.SIG.=  0.100D+03
         NO PREDICTION POSSIBLE


         ANALYSIS OF GROUP NO.  9          PREVIOUS GROUP NO.  3
NO.         NAME                 MC-SIG(%)   BON-F-SIG(%)      GROUPING
1           LANGUAGE             0.100D+03    0.100D+03        1   2
2           SEX                  0.100D+03    0.100D+03        1
3           NUMBER SERIES        0.774D+00    0.166D+00        1   2   /   3   4
4           FIGURE ANALOGIES     0.387D+00    0.749D-01        1   /   2   3   4
5           VERBAL REASONING     0.100D+03    0.100D+03        2


         BEST PREDICTOR
4           FIGURE ANALOGIES     0.387D+00    0.749D-01        1   /   2   3   4


STATISTICS FOR GROUPING:
GROUP NO.    MEAN   SAMPLE VARIANCE   NO.OF CASES    WEIGHT    WEIGHTED MEAN   VARIANCE(WEIGHTED MEAN)

    19     15.7534     8.8215           146       146.0000      15.7534            0.0604

    20     16.8571     9.4599           357       357.0000      16.8571            0.0265


         ANALYSIS OF VARIANCE:
                    SUM OF SQUARES     MEAN SQUARE      DEGREES OF FREEDOM
         GROUPING       126.2321         126.2321               1
         ERROR         4646.8378           9.2751             501

         F-VALUE    =    13.610
         SIGNIFICANCE=  0.250D-01
         BON-F-SIG.  =  0.749D-01
         MC-SIG.     =  0.387D+00
         CONSERV.SIG.=  0.749D-01
         GROUPING IS SIGNIFICANT AT THE   0.749D-01%- LEVEL(CONSERVATIVE)

         ANALYSIS OF GROUP NO. 10          PREVIOUS GROUP NO.  3
NO.         NAME                 MC-SIG(%)   BON-F-SIG(%)      GROUPING
1           LANGUAGE             0.994D+00    0.298D+01        1   /   2
2           SEX                  0.100D+03    0.100D+03        2
3           NUMBER SERIES        0.144D+01    0.341D+00        1   2   /   3   4
4           FIGURE ANALOGIES     0.669D+00    0.140D+00        1   2   3   /   4
5           VERBAL REASONING     0.100D+03    0.100D+03        2


         BEST PREDICTOR
4           FIGURE ANALOGIES     0.669D+00    0.140D+00        1   2   3   /   4


STATISTICS FOR GROUPING:
GROUP NO.    MEAN   SAMPLE VARIANCE   NO.OF CASES    WEIGHT    WEIGHTED MEAN   VARIANCE(WEIGHTED MEAN)

    21     17.9019     9.4475           316       316.0000      17.9019            0.0299

    22     19.7250     9.5378            40        40.0000      19.7250            0.2384


         ANALYSIS OF VARIANCE:
                    SUM OF SQUARES     MEAN SQUARE      DEGREES OF FREEDOM
         GROUPING       118.0099         118.0099               1
         ERROR         3347.9339           9.4574             354

         F-VALUE    =    12.478
         SIGNIFICANCE=  0.466D-01
         BON-F-SIG.  =  0.140D+00
         MC-SIG.     =  0.669D+00
         CONSERV.SIG.=  0.140D+00
         GROUPING IS SIGNIFICANT AT THE   0.140D+00%- LEVEL(CONSERVATIVE)

         ANALYSIS OF GROUP NO. 11          PREVIOUS GROUP NO.  4
NO.         NAME                 MC-SIG(%)   BON-F-SIG(%)      GROUPING
1           LANGUAGE             0.304D+01    0.911D+01        1   /   2
2           SEX                  0.100D+03    0.100D+03        1
3           NUMBER SERIES        0.132D+02    0.529D+01        1   /   2   3   4
4           FIGURE ANALOGIES     0.100D+03    0.100D+03        1   2   3   4
5           VERBAL REASONING     0.100D+03    0.100D+03        3


         BEST PREDICTOR
1           LANGUAGE             0.304D+01    0.911D+01        1   /   2


STATISTICS FOR GROUPING:
GROUP NO.    MEAN   SAMPLE VARIANCE   NO.OF CASES    WEIGHT    WEIGHTED MEAN   VARIANCE(WEIGHTED MEAN)

    23     17.6464     8.8298           181       181.0000      17.6464            0.0488

    24     18.3889     8.3996           126       126.0000      18.3889            0.0667


         ANALYSIS OF VARIANCE:
                    SUM OF SQUARES     MEAN SQUARE      DEGREES OF FREEDOM
         GROUPING        40.9525          40.9525               1
         ERROR         2639.3147           8.6535             305

         F-VALUE    =     4.7325
         SIGNIFICANCE=  0.304D+01
         BON-F-SIG.  =  0.911D+01
         MC-SIG.     =  0.304D+01
         CONSERV.SIG.=  0.304D+01
         NO PREDICTION POSSIBLE
```

```
         ANALYSIS OF GROUP NO. 12           PREVIOUS GROUP NO.  4
NO.          NAME                 MC-SIG(%)    BON-F-SIG(%)      GROUPING
 1        LANGUAGE                0.100D+03     0.100D+03        2   1
 2        SEX                     0.100D+03     0.100D+03        2
 3        NUMBER SERIES           0.100D+03     0.100D+03        1   2   3   4
 4        FIGURE ANALOGIES        0.394D+00     0.755D-01        1   2   /   3   4
 5        VERBAL REASONING        0.100D+03     0.100D+03        3


    BEST PREDICTOR
 4        FIGURE ANALOGIES        0.394D+00     0.755D-01        1   2   /   3   4


STATISTICS FOR GROUPING:
GROUP NO.    MEAN   SAMPLE VARIANCE   NO.OF CASES    WEIGHT    WEIGHTED MEAN   VARIANCE(WEIGHTED MEAN)

   23      18.5504    12.0463            129       129.0000      18.5504           0.0934

   24      19.9225     5.5721            129       129.0000      19.9225           0.0432


    ANALYSIS OF VARIANCE:
                       SUM OF SQUARES     MEAN SQUARE      DEGREES OF FREEDOM
    GROUPING               121.4302        121.4302               1
    ERROR                 2255.1474          8.8092             256

    F-VALUE      =     13.785
    SIGNIFICANCE=  0.252D-01
    BON-F-SIG.   =  0.755D-01
    MC-SIG.      =  0.394D+00
    CONSERV.SIG. =  0.755D-01
    GROUPING IS SIGNIFICANT AT THE   0.755D-01%- LEVEL(CONSERVATIVE)

         ANALYSIS OF GROUP NO. 13           PREVIOUS GROUP NO.  5
NO.          NAME                 MC-SIG(%)    BON-F-SIG(%)      GROUPING
 1        LANGUAGE                0.144D+00     0.431D+00        1   /   2
 2        SEX                     0.188D-01     0.565D-01        1   /   2
 3        NUMBER SERIES           0.715D+01     0.430D+01        1   2   /   3
 4        FIGURE ANALOGIES        0.785D+01     0.270D+01        1   2   /   3   4
 5        VERBAL REASONING        0.100D+03     0.100D+03        4


    BEST PREDICTOR
 2        SEX                     0.188D-01     0.565D-01        1   /   2


STATISTICS FOR GROUPING:
GROUP NO.    MEAN   SAMPLE VARIANCE   NO.OF CASES    WEIGHT    WEIGHTED MEAN   VARIANCE(WEIGHTED MEAN)

   25      19.3278     9.5065            180       180.0000      19.3278           0.0528

   26      20.6160     7.2062            125       125.0000      20.6160           0.0576


    ANALYSIS OF VARIANCE:
                       SUM OF SQUARES     MEAN SQUARE      DEGREES OF FREEDOM
    GROUPING               122.4234        122.4234               1
    ERROR                 2595.2298          8.5651             303

    F-VALUE      =     14.293
    SIGNIFICANCE=  0.188D-01
    BON-F-SIG.   =  0.565D-01
    MC-SIG.      =  0.188D-01
    CONSERV.SIG. =  0.188D-01
    GROUPING IS SIGNIFICANT AT THE   0.188D-01%- LEVEL(CONSERVATIVE)
```

From Figure 8.4 it appears that verbal reasoning (at the first partitioning level) contributes the most to the explanation of the variation in Y. The variables sex and number series appear at the second partitioning level. The categories of number series on the second level split in subgroups according to sex and the categories of sex on the second level split according to figure analogies. Higher order interactions are thus present and it is evident that a prediction model should perhaps rather be formulated per sex group.

6. General

For further theoretical details of the program CHAID, the reader may consult KASS (1980) and also the chapter on AID techniques in Hawkins (1982). The interpretation of the dendrogram is described in more detail by Muircheartaigh and Payne (1977).

A full description of the theoretical principles on which XAID is based, is given by Hawkins (1982).

When performing an AID analysis it often turns out that more than one predictor is highly significant. The AID procedure will split on that one which is the most significant. Suppose the highly significant predictors in ques-

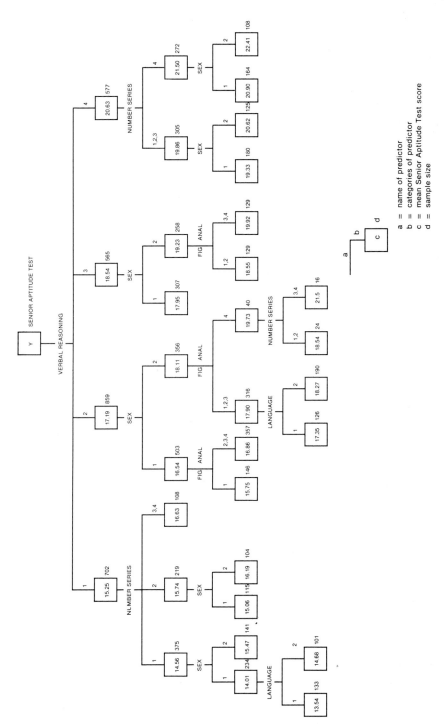

Figure 8.4. Dendrogram for the Senior Aptitude Test.

tion are language-group, population-group and socio-economic status. The researcher may consider socio-economic status as being a more crucial predictor than the other two. However, language-group may be slightly more significant and hence, the sample will be divided into subsamples according to the categories of language-group. In such a case, it may be preferable to first split the sample into subsamples according to the categories of socio-economic status and to then perform an AID analysis on each of the subsamples.

Magidson (1982) discusses some common pitfalls in the causal interpretation of the results of a categorical data analysis. He points out that a limitation of the AID approach is its inability to identify certain types of interactions.

CHAPTER 9

Control Charts

1. Introduction

An important aim of modern manufacturing processes is to produce articles that differ as little as possible with regard to certain important characteristics. It is interesting however, that despite the greatest care and precision, items produced still tend to vary owing to the combined effect of a variety of factors inherent in a manufacturing process.

It is customary to specify beforehand the permissible variation in the characteristics of items that are manufactured in order to ensure their usability. Items which do not meet the specified requirements should therefore, wherever possible, be eliminated prior to being distributed for use. In practice it frequently happens that specific assignable factors may influence the variation in the characteristics of an item. Examples of such factors are: reduced efficiency of the apparatus through wear and tear, a reduction in the voltage and the appointment of a new operator. Control charts are used in mass-production processes in order to indicate the presence of assignable sources of variation. If such a source is detected the process can be suspended temporarily while the source is eliminated.

The theory of control charts is based on the basic principle of stochastic variation. Plotted against time, stochastic variation in a process will indicate a completely random pattern. When the plot does not show a random pattern, this points to the presence of assignable factors in the process which increase the variation to such an extent that many (if not all) the manufactured items become unusable.

Each of a product's characteristics can be regarded as a variable which may be measured in either of the following ways:

(a) *Quantitative measurements*: The values of such a variable are usually of a continuous nature and include mass (in kilograms), diameter (in centimetres), volume (in litres), etc.

(b) *Dichotomous measurements*: The outcomes of such a variable determine whether or not the products meet the required standards, and are usually indicated as defective or non-defective. The characteristic in this case is the quality, which is measured as the ratio of defective items. Individual items therefore can not be assessed according to this characteristic. The quality is usually evaluated on the basis of a relatively large sample of items.

If a production process is "under control" (i.e., only stochastic variation is present) it is usually assumed that quantitative measurements of characteristics are normally distributed and that dichotomous measurements represent Bernoulli trials. This implies that the number of defective items in a group are binomially distributed. If the probability of obtaining a defective item is smaller than 0.10 and large groups of items are examined, the Poisson model may be used as an approximation of the binomial model.

2. Process Capability

It has already been stated that items which can be classified only as defective or nondefective are not examined individually, but in samples. Although this is not essential, items with quantitative characteristics are usually also examined in samples instead of individually. The following table gives some guidelines in respect of sample size.

Nature of characteristic	Limits for sample sizes	Sample sizes generally used
Dichotomous	$25 \leq n \leq 250$	50, 100 and 200
Quantitative	$2 \leq n \leq 12$	4 and 5

Suppose m samples of size n each are drawn at regular intervals (for instance hourly or daily) in the course of the production process and the relevant measurements are carried out on each of the mn items.

In the case of quantitative measurements the sample means and range can be calculated for each sample and schematically represented as follows:

Sample means: $\bar{X}_1, \bar{X}_2, \ldots, \bar{X}_m$
Sample range: $W_1, W_2, \ldots, W_m$.

The process capability is now defined as

$$S_0 = \bar{W}/d_n$$

Table 9.1. Hartley's conversion
constant for different values
of n

Sample size n	Conversion constant d_n
2	1.128
3	1.693
4	2.059
5	2.326
6	2.534
7	2.704
8	2.847
9	2.970
10	3.078
11	3.173
12	3.258

where

$$\overline{W} = (1/m) \sum_{i=1}^{m} W_i$$

and d_n is known as Hartley's conversion constant. Values of d_n [Murdoch (1979)] for $n = 2$ to $n = 12$ are given in Table 9.1.

The process capability is used for determining the limits of control charts and can be regarded as a measure of the typical stochastic variation in the process.

For dichotomous measurements the process capability is designated as P_0 and is given by:

$$P_0 = (\text{Total number of defectives in all } m \text{ samples})/mn.$$

3. Control Charts for Items with Quantitative Characteristics

A control chart consists of two sets of limits, namely the warning limits and the action limits. For quantitative characteristics two types of control charts are usually compiled, namely $\overline{X}$-charts and W-charts.

A. Arithmetic Mean Control Charts ($\overline{X}$-Charts)

Means $\overline{X}_1, \overline{X}_2, \ldots, \overline{X}_m$ of the m samples of size n each are calculated and plotted on the axial system of the control chart in the order of their collection. If the process mean is given by

$$\bar{X} = (1/m) \sum_{i=1}^{m} X_i,$$

the control chart limits are determined as follows:

Warning limits: $\bar{X} \pm 1.96 S_0/\sqrt{n}$ (approximately $\bar{X} \pm 2 S_0/\sqrt{n}$)
Action limits: $\bar{X} \pm 3.09 S_0/\sqrt{n}$ (approximately $\bar{X} \pm 3 S_0/\sqrt{n}$).

If the process is under statistical control, there is a 0.05 probability that an observed sample mean will lie outside the warning limits and a 0.002 probability that it will be outside the action limits.

B. Range Control Charts (W-Charts)

In addition to the control chart for the sample means a further control chart is usually compiled in order to determine whether the variation in each sample is not becoming excessive. This variation is not shown by the mean of a sample. Since it should only be ensured that this variation does not get out of hand, only upper limits are determined. These limits are the following:

Upper warning limit: $\overline{W} D_w$
Upper action limit: $\overline{W} D_a$.

The probability that an observed W-value is larger that the upper warning limit and action limit is respectively 0.025 and 0.001. Values of D_w and D_a can be read from Table 9.2. These values have been obtained from Murdoch (1979).

The constants given in Tables 9.1 and 9.2 were determined on the assump-

Table 9.2. Control chart constants for the W-chart

Sample size	Constants for upper limit	
n	D_w	D_a
2	2.81	4.12
3	2.17	2.98
4	1.93	2.57
5	1.81	2.34
6	1.72	2.21
7	1.66	2.11
8	1.62	2.04
9	1.58	1.99
10	1.56	1.93
11	1.53	1.91
12	1.51	1.87

tion that the measurements of the characteristics under discussion can be regarded as a sample from a normal population.

EXAMPLE 9.1. According to the specifications the diameter of a particular car spare part should be 1 cm. In order to determine whether the production process of that part is under statistical control, four newly manufactured parts are examined every hour. The data of 25 successive samples are given in the table below.

Sample	$\bar{X}_i$	W_i	Sample	$\bar{X}_i$	W_i
1	1.002	0.008	14	0.995	0.032
2	1.002	0.002	15	0.997	0.016
3	0.992	0.038	16	0.998	0.009
4	1.005	0.025	17	1.000	0.018
5	0.995	0.012	18	1.005	0.023
6	1.004	0.020	19	1.002	0.019
7	1.001	0.007	20	1.000	0.031
8	0.998	0.019	21	1.001	0.024
9	1.003	0.004	22	0.996	0.015
10	1.001	0.015	23	0.998	0.007
11	0.999	0.013	24	0.997	0.033
12	1.004	0.011	25	0.999	0.008
13	1.006	0.022			

Now $\overline{W} = (1/25)(0.008 + 0.002 + \cdots + 0.008) = 0.017$ cm.

From Table 9.1 it follows that $d_4 = 2.059$ and consequently the process capability is:

$$S_0 = \overline{W}/d_4$$

$$= 0.017/2.059$$

$$= 0.008.$$

The process mean equals:

$$\bar{X} = (1/25) \sum_{i=1}^{25} \bar{X}_i$$

$$= (1/25)(1.002 + 1.002 + \cdots + 0.999) = 1.$$

$\bar{X}$-Chart.

From the above it follows that the warning limits and action limits are respectively given by:

Warning limits: $1 \pm 1.96 \times 0.008/\sqrt{4}$
                   $= 1 \pm 0.0078$ cm
Action limits: $1 \pm 3.09 \times 0.008/\sqrt{4}$
                   $= 1 \pm 0.0124$ cm.

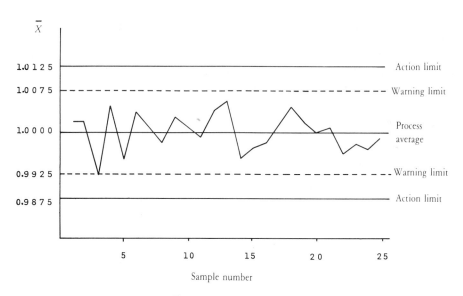

Figure 9.1. $\bar{X}$-chart for the data of Example 9.1.

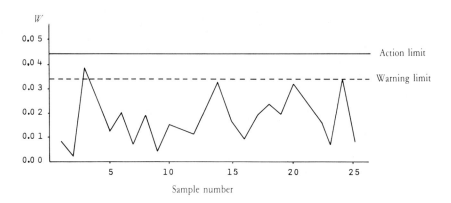

Figure 9.2. W-chart for the data of Example 9.1.

Figure 9.1 shows that $\bar{X}$-control chart based on these obtained limits.

W-Chart.

    Upper warning limit: $\overline{W}D_w$
$$= 0.017 \times 1.93$$
$$= 0.0328$$

    Upper action limit: $\overline{W}D_a$
$$= 0.017 \times 2.57$$
$$= 0.0437.$$

Figure 9.2 is a graphical representation of the W-chart.

The $\bar{X}$ and W-charts show that the process was under statistical control during the 25 hours of production—only Sample 3 produced $\bar{X}$ and W values falling outside the warning limit. Since approximately 1/20 of all means are expected to fall outside the limits, even if the process is under control, this one value falling outside the warning limit should therefore be no cause for concern.

In order to obtain the maximum benefit from a control chart, it would be meaningful to determine the limits for observed samples of items while the process is under control. New $\bar{X}_i$ or W_i-values can then be plotted sequentially and evaluated with a view to suspending the production process if the values systematically fall outside the action limits.

After the samples have been plotted, the control chart is usually examined in order to determine whether all the points lie inside the limits. Points falling outside the limits are eliminated, after which a new control chart is compiled.

4. Control Charts for Dichotomous Measurements (*P*-Chart)

In this case the warning and action limits are determined by means of the normal approximation of the binomial distribution when the process capability $P_0 \geq 0.10$, and by means of the Poisson approximation of the binomial distribution when $P_0 < 0.10$.

The limits for the ratio of defective items in a sample of size n are:

(i) $P_0 \geq 0.10$

Warning limits: $nP_0 \pm 1.96\sqrt{nP_0(1 - P_0)}$

Action limits: $nP_0 \pm 3.09\sqrt{nP_0(1 - P_0)}$

If a negative value is obtained for the lower limit, a lower limit value of zero is taken.

(ii) $P_0 < 0.10$

Usually only upper limits are determined in the case of $P_0 < 0.10$. Let λ indicate the parameter of the Poisson distribution. An estimate of λ is nP_0.

Designate the upper warning and action limits respectively as C_w and C_a. First determine the smallest value of C_w for which $P(X \geq C_w) \leq 0.025$, where X follows the Poisson (λ) distribution. Subsequently determine the smallest value of C_a for which $P(X \geq C_a) \leq 0.001$. These numbers can be obtained from a table of Poisson probabilities by substituting $\lambda \cong nP_0$.

EXAMPLE 9.2. A government institution wishes to check the accuracy of the work produced by their typing pool. To begin with, samples are taken of the work of all the typists in order to compile a *P*-chart. Next, the obtained control limits are used for running a quality check on the work of each typist individually. If a typed page contains two or more errors, this page is designated as one that is defective. A sample of 50 typed pages is drawn daily and the number of "defective" pages determined. The number of defectives in 25 successive samples is as follows:

Sample no.	1	2	3	4	5	6	7	8	9	10	11	12	13
Number of defectives	2	2	5	6	3	5	2	1	1	0	0	1	0

Sample no.	14	15	16	17	18	19	20	21	22	23	24	25
Number of defectives	1	0	2	1	0	0	1	1	0	0	1	0

$$\text{The process capability } P_0 = \frac{\text{total number of defectives}}{\text{number of samples} \times \text{sample size}}$$

$$= \frac{35}{1250} = 0.028.$$

Since $P_0 < 0.10$ the Poisson distribution is used. The mean number of defectives per sample of size 50 is $\lambda = nP_0 = 50 \times 0.028 = 1.4$.

The upper warning limit is obtained as the smallest value C_w, for which $P(X \geq C_w) \leq 0.025$ where X follows a Poisson distribution with $\lambda = 1.4$. From a table of Poisson probabilities it follows that $P(X \geq 4) = 0.0537$ and $P(X \geq 5) = 0.0143$. Since the probability 0.0143 is the nearest to 0.025, $C_w = 5$ is used.

The upper action limit is the smallest value C_a for which $P(X \geq C_a) \leq 0.001$. From a table of Poisson probabilities with $\lambda = 1.4$ it follows that $P(X \geq 6) = 0.0032$ and $P(X \geq 7) = 0.0006$ so that $C_a = 7$.

Inspection shows that the number of defectives of Samples 3, 4, and 6 is larger than or equivalent to the warning limit. On closer examination it appears that the assignable source of variation present in Samples 1 to 6 had been eliminated prior to taking Sample 7. (One of the typists concerned had

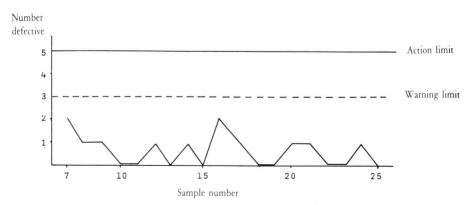

Figure 9.3. *P*-chart for the data of Example 9.2.

resigned). Beginning with Sample 7 it follows that the process capability is now $P_0 = 12/950 = 0.0126$.

Consequently $\lambda = nP_0 = 50 \times 0.0126 = 0.63$. For $\lambda = 0.6$ it follows that $C_w = 3$ and $C_a = 5$. The P-chart is represented graphically in Figure 9.3.

5. Cumulative Sum Charts

In the case of small volume production processes cumulative sum (CUSUM) charts, where samples of size $n = 1$ are drawn each time, are an ideal mechanism for checking statistical control. CUSUM charts are also much more effective in detecting small changes in the production process than the usual control charts.

If items are examined at regular intervals and the measurements of a particular characteristic are represented by $X_1, X_2, \ldots, X_m$, a CUSUM chart is compiled as follows: Plot the cumulative sums $S_1, S_2, \ldots, S_m$ against the collection order $1, 2, \ldots, m$ where

$$S_1 = (X_1 - k)$$
$$S_2 = (X_1 - k) + (X_2 - k) = S_1 + (X_2 - k)$$
$$\vdots$$
$$S_m = \sum_{i=1}^{m} (X_i - k) = S_{m-1} + (X_m - k)$$

and k is a constant reference value. Usually k is taken as the process mean

$$\bar{X} = (1/m) \sum_{i=1}^{m} X_i$$

derived from a previous data set.

A generally ascending trend in the CUSUM graph between two time units indicates an ascending trend in the process mean in the corresponding period. A descending trend in the CUSUM graph, on the other hand, indicates a decrease in the process mean. If the CUSUM graph stays relatively constant, this indicates that there is no change in the mean, i.e. the process is under statistical control. A change in the process mean is therefore characterized by a change in the slope of the CUSUM graph.

In practice stochastic fluctuation will give rise to small fluctuations in the slope of the CUSUM graph. Consequently the mean slope is usually determined at a number of consecutive points in order to ascertain whether there has been a change in the level of the mean over a specific period.

EXAMPLE 9.3. The director of a computer centre reports the following number of system interruptions per week over a period of 50 weeks.

Week no.	Number of system interruptions per week				
1– 5	1	4	6	5	1
6–10	5	0	3	1	1
11–15	3	2	1	5	2
16–20	5	3	4	3	4
21–25	3	1	4	1	4
26–30	0	0	4	0	2
31–35	2	5	1	2	3
36–40	4	1	2	0	0
41–45	5	1	5	5	4
46–50	0	5	4	4	2

Figure 9.4 is a graphical representation of the data.

From previous experience the director knows that 3 interruptions occur on the average per week. Therefore $k = 3$ is taken as a convenient constant reference value. The CUSUM values $S_m = \sum_{i=1}^{m} (X_i - 3)$, $m = 1, 2, \ldots, 50$ are given in the table below and represented graphically in Figure 9.5.

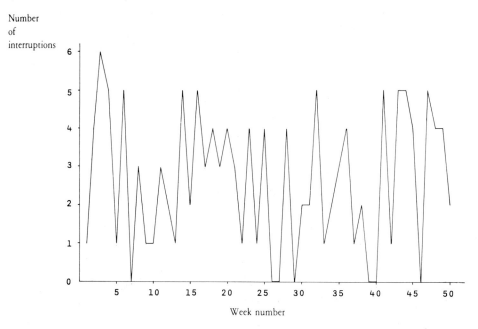

Figure 9.4. Number of system interruptions per week.

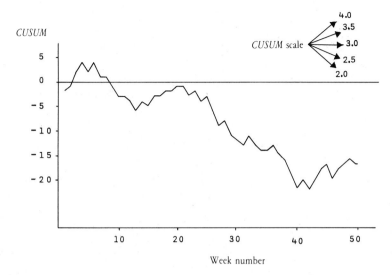

Figure 9.5. CUSUM chart for the number of system interruptions at a computer center.

Sample no.	CUSUM values				
1– 5	−2	−1	2	4	2
6–10	4	1	1	−1	−3
11–15	−3	−4	−6	−4	−5
16–20	−3	−3	−2	−2	−1
21–25	−1	−3	−2	−4	−3
26–30	−6	−9	−8	−11	−12
31–35	−13	−11	−13	−14	−14
36–40	−13	−15	−16	−19	−22
41–45	−20	−22	−20	−18	−17
46–50	−20	−18	−17	−16	−17

It is customary to provide a CUSUM scale with the CUSUM chart. This scale gives the slope of the graph with regard to the reference value. To illustrate this, suppose the following axial system is used:

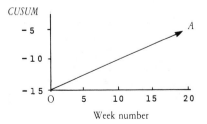

From this representation it follows that the slope of line segment OA equals $(-5 - (-15))/(20 - 0) = 0.5$. This value for the slope indicates a mean increase of 0.5 relative to the k value. In Figure 9.5 the line segment OA will have a CUSUM scale value of 3.5, since the reference value $k = 3$. If a section of the graph, for instance, corresponds to a slope of 1.0 on the CUSUM scale, this value corresponds to an average of one system interruption per week for the period under consideration. With reference to Figure 9.5 it therefore follows that:

During the first five weeks the number of system interruptions averaged approximately 4.0 a week. The following table shows the approximate number of interruptions a week for the rest of the period:

Weeks 6–12	2.0 a week
Weeks 13–20	3.5 a week
Weeks 21–40	2.0 a week
Weeks 41–50	3.5 a week

The above example clearly illustrates the ability of a CUSUM control chart to point out systematic changes that occur.

EXAMPLE 9.4. The table below gives the percentage of the potential listeners who tuned in to a particular radio serial during the first 38 weeks of broadcast.

Week no.	%	Week no.	%	Week no.	%
1	13.5	14	9.4	27	6.3
2	16.8	15	15.6	28	9.5
3	12.5	16	6.7	29	—
4	15.6	17	7.8	30	—
5	—	18	8.5	31	7.0
6	9.9	19	9.6	32	6.4
7	24.9	20	—	33	3.5
8	16.2	21	7.9	34	5.8
9	21.6	22	13.4	35	7.9
10	—	23	13.0	36	6.3
11	9.2	24	7.3	37	5.9
12	8.5	25	5.8	38	5.1
13	7.0	26	10.1		

Five missing values occur in the table above. The mean for the remaining values is 10.14%. Since no prior information is available we take $k = 10\%$ as a convenient reference value. Note that missing data are replaced with the reference value k, which means that the CUSUM remains constant at the value which precedes the missing data. In some cases it may be more accurate to replace each missing value by the mean of its neighbouring values. The CUSUM values for compiling the CUSUM graph are as follows:

Week no.	CUSUM	Week no.	CUSUM	Week no.	CUSUM
1	3.5	14	45.1	27	37.1
2	10.3	15	50.7	28	36.6
3	12.8	16	47.4	29	36.6
4	18.4	17	45.2	30	36.6
5	18.4	18	43.7	31	33.6
6	18.3	19	43.3	32	30.0
7	33.2	20	43.3	33	23.5
8	39.4	21	41.2	34	19.3
9	51.0	22	44.6	35	17.2
10	51.0	23	47.6	36	13.5
11	50.2	24	44.9	37	9.4
12	48.7	25	40.7	38	4.5
13	45.7	26	40.8		

The CUSUM chart is given in Figure 9.6, from which it appears that the following changes occurred over the period of observation:

Weeks 1–10: 13% on average tuned in to the program
Weeks 11–23: 10% on average tuned in to the program
Weeks 24–38: 8% on average tuned in to the program.

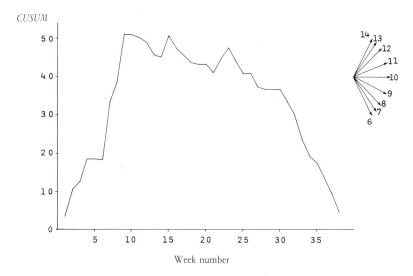

Figure 9.6. CUSUM chart for the percentage of persons who tuned in to a particular radio serial.

From the above data it appears that the above serial's popularity began to wane after about 23 weeks.

EXAMPLE 9.5. The number of road accidents per year and the reasons for these accidents continue to be a matter of great concern to the authorities in the RSA. During the period 1960 to 1966 no speed limit was imposed on the national highways of the Republic of South Africa. From 1967 to 1973 a speed limit of 120 km/h was imposed, but due to the oilshortage this was lowered to 90 km/h during the period 1974 to 1980 and then increased again to 100 km/h for the years 1981 to 1982.

A CUSUM chart of the number of fatal casualties per 10,000 vehicles on the road may be used in an exploratory analysis to detect whether the number of deaths due to road accidents is related to the speed limit. The relevant road casualties data is given as part of the next program for calculating and then plotting the CUSUM-values.

Input: SAS PROC MATRIX program and SAS PROC GPLOT program for obtaining a CUSUM chart

```
PROC    MATRIX ;
* Y = NUMBER OF FATAL ROAD ACCIDENTS PER 10000 VEHICLES ;
Y =     24.66 29.86 30.93 34.71 36.41 37.01 34.45 34.88 33.62 36.24
        38.62 36.48 36.17 32.77 22.45 26.95 25.08 20.04 19.78 17.95
        21.67 24.30 22.57 ;
* X = CORRESPONDING YEAR ;
X =     1960 1961 1962 1963 1964 1965 1966 1967 1968 1969 1970 1971
        1972 1973 1974 1975 1976 1977 1978 1979 1980 1981 1982  ;
        NOBS = NCOL(Y) ;
        REF = SUM(Y)#/NOBS ;    * COMPUTE MEAN NUMBER OF ACCIDENTS ;
NOTE    REFERENCE VALUE ;
PRINT   REF ;
        CUSUM = J(NOBS,1,0) ;   * COMPUTE CUSUM ;
        SUMY = 0 ;
DO K =  1 TO NOBS ;
        SUMY = SUMY + (Y(,K)-REF) ;
        CUSUM(K,) = SUMY ;
END ;
        SUMM = X'||CUSUM ;
OUTPUT SUMM DATA=CUSUMS (RENAME=(COL1=YEAR COL2=CUSUM )) ;
PROC    PRINT DATA=CUSUMS ;
PROC    GPLOT ;
        PLOT  CUSUM*YEAR=1 /  HAXIS = 1955 TO 1985 BY 10
                              VAXIS = -5 TO 65 BY 10 ;
GOPTIONS HSIZE=7 VSIZE=7 SPEED=7 ;
        SYMBOL1 I=JOIN  C=BLACK ;
FOOTNOTE .C=BLACK 'Number of fatal casualties per 10000 vehicles' ;
```

Output:

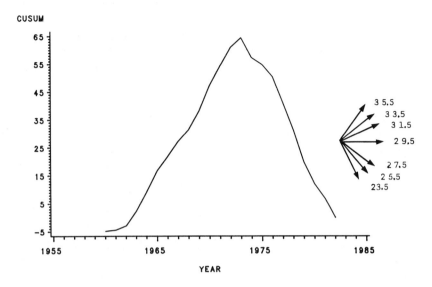

Number of fatal casualties per 10000 vehicles

It appears from the CUSUM chart that the following changes occured over the period of observation:

1960–1962: Approximately 30 accidents per 10,000 vehicles
1963–1973: Approximately 35.5 accidents per 10,000 vehicles
1974–1982: Approximately 23 accidents per 10,000 vehicles.

The average number of accidents during the years 1960–1982 is 29.5 and this figure was taken as the reference value. From this analysis it seems that a decrease in the number of fatal accidents accompanied the lowering of the speed limit.

6. Cumulative Sine Charts

The cumulative sine (CUSIN) chart is also a visual representation of data collected in a particular order. The CUSIN chart can be compared to the CUSUM chart, but differs from it in that a change in the process mean is indicated by a change in angle direction (angle size) instead of a change in slope.

In quality control it is standard procedure to make adjustments to the production apparatus when a change in the production mean is detected and then to start with a new CUSUM chart. If in this case the old CUSUM chart is retained however, it becomes increasingly difficult to detect changes in the process by visually inspecting the CUSUM chart.

Situations arise in which it is not practically possible to interrupt the production process or where all that is needed is to make a brief study of changes in the whole observed series. Although the CUSIN chart was originally developed as a control chart, it is more appropriate for analyzing a process retrospectively.

For convenience sake the initial mean is represented as a horizontal line. Angles with regard to this line are measured anticlockwise according to convention. The angle between a particular section of the CUSIN and the horizontal is proportional to the deviation from the process mean. Suppose a specific car part has a specified diameter of 100 mm and take as angle scale $3°$ for every 1 mm deviation from 100 mm.

A diameter of 115 mm is equivalent in this case to an angle direction of $45°$, a diameter of 80 mm to an angle direction of $-60°$ and a diameter of 102 mm to an angle direction of $6°$. These deviations from the specified diameter can be represented graphically as follows:

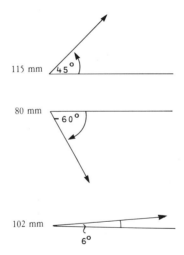

The form of the CUSIN chart is determined by the choice of an angle scale. As general guideline the angle scale is chosen in such a way that $30°$ correspond roughly to one standard deviation of the observed deviations [North (1980)]. The CUSIN chart can then be obtained as follows:

Suppose a line segment has length a and that the angle scale is θ for a given number of deviations from the process mean (as represented graphically in the figure below).

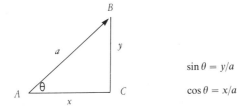

$$\sin \theta = y/a$$
$$\cos \theta = x/a$$

From this figure it follows that the y co-ordinate is given by $y = a \sin \theta$ whereas the x co-ordinate is given by $a \cos \theta$. The CUSIN chart is obtained by plotting the points S_m against C_m, $m = 1, 2, \ldots$ where

$$C_m = a \sum_{i=1}^{m} \cos(\theta(X_i - k))$$

$$S_m = a \sum_{i=1}^{m} \sin(\theta(X_i - k)).$$

In the above notation X_i indicates the process observation, k the reference value, θ the angle scale and a the length of the line segment AB.

A CUSIN chart is compiled by means of a graphical plotter linked to the computer. Alternatively the CUSIN chart can be compiled by making use of a protractor.

EXAMPLE 9.6. Certain institutions use a flexitime system instead of fixed working hours for their personnel. The following table shows the flexitime credits that a particular person carried forward from one month to the next month over a period of 36 months.

Month no.	No. of hours' flexitime carried forward to following month					
1– 6	−1.10	1.62	0.95	−0.85	3.71	0.18
7–12	−1.40	−2.99	2.44	0.81	−1.03	4.50
13–18	0.34	−0.26	−1.51	−1.21	−3.68	−0.98
19–24	−0.01	−1.15	0.55	−2.40	−0.71	−0.83
25–30	0.42	2.24	4.42	4.15	2.92	5.00
31–36	3.62	2.63	−0.82	−1.62	−2.54	0.20

The arithmetic mean and standard deviation are $\bar{X} = 0.434$ and $s = 2.29$. As constant reference value $k = 0$ is taken. Since $s = 2.29$ we take as angle scale $30° = 2$ hours (i.e., $15° = 1$ hour).

The co-ordinates C_m and S_m are subsequently calculated, for instance, by substituting $a = 0.5$ cm, $\theta = 15°$ and $k = 0$. From this it follows for $m = 1$, 2, ..., 36 that

$$C_m = 0.5 \sum_{i=1}^{m} \cos(15\ X_i)$$

$$S_m = 0.5 \sum_{i=1}^{m} \sin(15\ X_i); \qquad m = 1, 2, \ldots, 36.$$

As an alternative a CUSIN chart is compiled by means of a protractor. Since $X_1 = -1.10$ this deviation from a constant reference value $k = 0$ is indicated by an angle of $15° \times -1.10 = -16.5°$ and a line of length a cm. The angle directions corresponding to the second and third deviations are respectively $15° \times 1.62 = 24.3°$ and $15° \times 0.95 = 14.3°$. Construction of the CUSIN chart is then as follows.

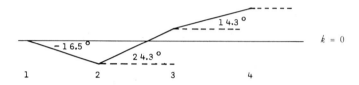

The program for calculating and then plotting the CUSIN values is subsequently given.

Input: SAS PROC MATRIX program and SAS PROC GPLOT program for obtaining a CUSIN chart

```
PROC    MATRIX ;
X =    -1.10  1.62   0.95 -0.85   3.71  0.18
       -1.40 -2.99   2.44  0.81  -1.03  4.50
        0.34 -0.26  -1.51 -1.21  -3.68 -0.98
       -0.01 -1.15   0.55 -2.40  -0.71 -0.83
        0.42  2.24   4.42  4.15   2.92  5.00
        3.62  2.63  -0.82 -1.62  -2.54  0.20  ;
        M = NCOL(X) ;
        CM = J(M,1,0) ;
        SM = J(M,1,0) ;
        CSUMX = 0.0 ;
        SSUMX = 0.0 ;
DO K = 1 TO M ;                         * Compute CUSIN ;
        RAD = 0.0174532 ;               * Convert degrees to radians ;
        DEGRAD = 15.0*RAD*X(1,K) ;
        CSUMX = CSUMX + COS(DEGRAD) ;
        SSUMX = SSUMX + SIN(DEGRAD) ;
        CM(K,1) = 0.5 * CSUMX ;
        SM(K,1) = 0.5 * SSUMX ;
END ;
        SUMM = CM||SM ;
OUTPUT SUMM DATA=CUSIN (RENAME=(COL1=CM COL2=SM )) ;
PROC    PRINT DATA=CUSIN ;
TITLE .C=BLACK CUSIN Plot ;
PROC    GPLOT ;
        PLOT   SM*CM=1 / VREF = 0 VAXIS = -1.0 TO 2.5 BY 0.5 ;
GOPTIONS HSIZE=7 VSIZE=7 SPEED=7 ;
        SYMBOL1 I=JOIN  C=BLACK ;
FOOTNOTE1 ;
FOOTNOTE2 .C=BLACK 'Flexitime data' ;
```

Output:

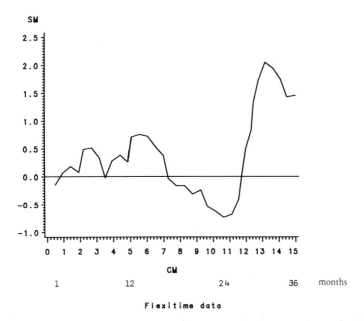

Flexitime data

From the CUSIN chart it appears that this particular person maintained fairly constant flexitime credits during the first year. During the second year there was a tendency to carry forward flexitime debits, whereas during the third year reasonably large credits were carried over to the next month. Note that the month numbers on the horizontal axis of a CUSIN chart will usually not be equally spaced.

7. General

For further details and practical applications of control charts and specifically how to perform statistical inference when constructing a CUSUM, the reader is referred to Murdoch (1979).

The quality of a product usually depends on several characteristics and it may be necessary to carry out a simultaneous control of these characteristics. Jackson (1959) discussed techniques for the quality control of several variables and Hotelling (1947) has suggested the use of the T^2-statistic in the use of multivariate quality control.

A procedure which is graphical in nature and which is based on the Andrews' function plot technique (see Chapter 4, Section 7) is described in an article by Kulkarni and Paranjape (1984) and illustrated by these authors with the aid of simulated data.

CHAPTER 10

Time Series Representations

1. Representations in the Time Domain

Time series data are usually collected on a monthly, quarterly or annual basis. Such time series provide important economic and demographic information and are published by various institutions on a regular basis.

Usually it is assumed that the time series values, Y, are given by the product of four movements, called components. In other words

$$Y = TCSI$$

where T indicates the trend or long-term movement, C the cyclical movement, S the seasonal variation and I irregular or stochastic variation.

Although the trend and seasonal variation are usually the most important, studying the cyclical variation and irregular variation frequently also gives rise to useful information.

It is customary to fit a curve to the data in order to obtain an estimate of the trend. The form of the curve can be determined if a moving average is constructed. The moving average series is obtained by substituting for every observed Y value the mean of this value and adjacent Y values.

Symbolically the calculation of a three-year moving average is represented as follows:

Year	Y_i	Three-year moving average
1970	Y_1	
1971	Y_2	$\ldots(Y_1 + Y_2 + Y_3)/3$
1972	Y_3	$\ldots(Y_2 + Y_3 + Y_4)/3$
1973	Y_4	$\ldots(Y_3 + Y_4 + Y_5)/3$
1974	Y_5	

Table 10.1. Total number of road accidents in the RSA: 1970–1979

Year	Y_i	Five-year moving total	Five-year moving average
1970	205,267		
1971	218,926		
1972	218,436	1,077,265	215,453.0
1973	221,528	1,132,760	226,552.0
1974	213,108	1,183,889	236,777.8
1975	260,762	1,228,818	245,763.6
1976	269,565	1,295,928	259,185.6
1977	263,365	1,347,919	269,583.8
1978	288,638		
1979	265,099		

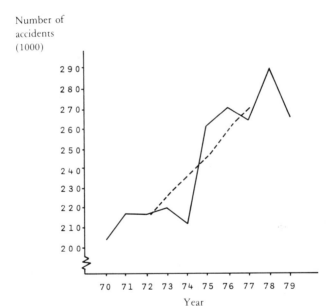

Number of
accidents
(1000)

Figure 10.1. Annual number of road accidents in the RSA (the dotted line indicates the five-year moving average).

Table 10.1 gives the total number of road accidents in the Republic of South Africa (RSA) for the period 1970–1979 obtained from the Central Statistical Service of the RSA. Figure 10.1 is a graphical representation of the Y values as well as the five-year moving average.

Note that the successive points in Figure 10.1 are joined by straight lines in contrast with the scatterplots in regression analysis. The five-year moving average shows that the trend may be described adequately by a straight line.

Table 10.2. Specific seasonal indices for the value of RSA wool exports: 1975–1979

Month	1975	1976	1977	1978	1979	Typical seasonal index
Jan.		71.98	81.16	88.64	78.02	79.95
Feb.		119.26	117.32	104.87	121.58	115.76
March.		151.65	165.92	113.56	129.73	140.22
Apr.		121.85	91.46	119.27		110.86
May		102.88	119.17	94.72		105.59
June		116.47	113.99	102.27		110.91
July	62.02	51.40	40.91	58.82		53.29
Aug.	32.84	49.54	40.34	57.31		45.00
Sept.	55.51	71.27	73.75	91.41		72.99
Oct.	95.07	121.76	130.65	118.06		116.39
Nov.	138.12	113.74	117.36	123.95		123.29
Dec.	137.06	111.23	140.98	115.59		126.22

When data are collected on a monthly basis, some time series indicate a pattern of variation which is repeated every 12 months. In this case a 12-month moving average produces a relatively reliable estimate of the TC component. This fact is used for isolating the SI component through division of the original Y values by the TC values. In other words

$$SI = \frac{TCSI}{TC}, \qquad \text{where } Y = TCSI.$$

Table 10.2 gives the specific seasonal indices $(SI \times 100)\%$ for the value of RSA wool exports for the period 1975 to 1979 as reported by the Central Statistical Service of the RSA.

By calculating the averages (or medians) for each month, the irregular variation is eliminated from SI, and a so-called typical seasonal index is then obtained. The final column of Table 10.2 gives the typical seasonal index for the value of wool exports. Figure 10.2 is a graphical representation of the specific seasonal indices for 1976–1978 and Figure 10.3 of the typical seasonal index.

From these figures it appears that the March activity is roughly 40% higher than the average annual activity. In contrast with this the August activity is approximately 55% lower than the average annual activity. Summarising, it can be stated that RSA wool exports are below average during January, July, August and September and above average in the remaining months.

Some institutions publish a so-called seasonally adjusted time series instead of the raw data. Such a series is obtained by dividing the original Y values by the typical seasonal index values S.

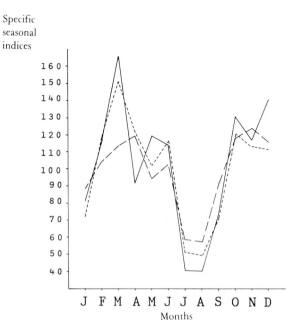

Figure 10.2. Specific seasonal indices for 1976–1978 for the value of RSA wool exports.

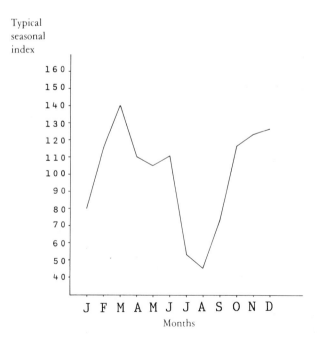

Figure 10.3. Typical seasonal indices for the value of RSA wool exports.

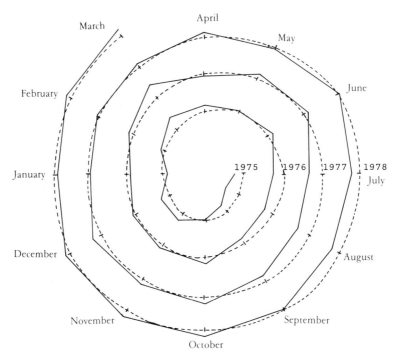

Figure 10.4. Seasonal cobweb for the value of RSA wool exports.

As alternative to Figure 10.2 the specific seasonal indices can also be represented in a seasonal cobweb or a seasonal needle diagram (see Figures 10.4 and 10.5). These figures are self-explanatory and also clearly show, in accordance with Figure 10.2, that a relatively similar seasonal pattern prevailed during 1975 to 1979.

It can sometimes also be useful to isolate the irregular movement and represent it graphically. Every I value is obtained by dividing the corresponding SI value by the typical seasonal index S for the particular month. The graphical representation of the I component is used for identifying possible outlying Y values as well as for confirming assumptions of distribution in respect of I.

Representing the Y, TC, SI, and I components of an extensive time series observed on a monthly basis graphically, gives rise to a global interpretation of the data [Cleveland and Terpenning (1982)]. In order to illustrate this, the next two time series are represented graphically by means of the following SAS PROC MATRIX and SAS PROC GPLOT programs:

Time Series 1: Monthly value of RSA wool exports (million rands) 1973–1982.

Time Series 2: Monthly Johannesburg Stock Exchange (JSE) index for coal shares 1970–1981.

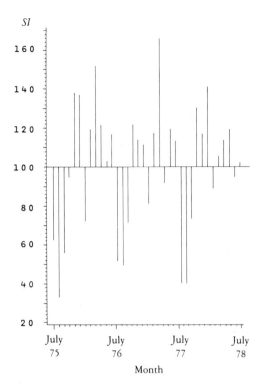

Figure 10.5. Seasonal needle diagram of the value of RSA wool exports.

Input: Time Series 1

```
PROC    MATRIX ;
        A =13.4 13.1 17.9 17.0 18.2 18.3  7.9   8.3  9.1 18.1 19.2 19.9
           12.5 15.8 15.8 12.9 15.1 12.6  6.3   4.8  1.6  7.4  9.9 12.7
            7.1  9.2 10.4 14.1 12.0 12.1  7.2   4.0  7.3 13.3 20.0 20.6
           11.1 18.7 24.4 20.3 17.5 19.8  8.8   8.6 12.6 21.6 20.2 20.0
           14.6 21.0 29.7 16.5 21.7 21.1  7.7   7.6 13.6 24.0 21.7 25.8
           16.3 19.6 21.6 22.9 18.3 19.8 11.3  11.1 18.1 23.6 25.0 23.5
           15.0 24.0 26.4 22.6 23.0 19.0 13.3  11.6 14.4 24.7 22.2 23.3
           23.2 30.6 31.8 21.4 19.6 16.9 11.3  13.5 17.9 24.3 22.5 28.1
           21.6 28.3 30.5 25.2 28.3 22.1 19.4  15.6 20.5 27.5 34.7 34.1
           24.5 34.0 42.5 29.0 29.8 28.2 17.8  15.7 20.9 20.9 25.3 30.0 ;
        Y= A' ;
* Y, THE TIME SERIES MUST BE COMPLETE I.E. EACH YEAR MUST HAVE
      DATA FOR 12 MONTHS ;
      ROWN = 'July' / 'Aug' / 'Sept' / 'Oct'/ 'Nov' / 'Dec' / 'Jan'/
             'Febr' / 'March' / 'April' / 'May' / 'June' ;
      NOBS = NROW(Y) ;            * NUMBER OF MONTHS ;
      NYRS = NOBS #/ 12 ;         * NUMBER OF YEARS ;
      NMOVE = NOBS - 11 ;
      NY1 = NYRS - 1 ;
      NGMOVE = NMOVE - 1 ;
      SUMM= J(NMOVE,1,0) ;          * INITIALIZE  SUMM;
      TC = J(NGMOVE,1,0) ;          * INITIALIZE TC ;
* CALCULATE THE CENTERED MOVING AVERAGES (TC), SPECIFIC
   SEASONAL INDICES (SI) AND THE IRREGULAR COMPONENTS (IR) ;
      YY = J(NGMOVE,1,0) ;
      IR = J(NGMOVE,1,0) ;
      SI = J(NGMOVE,1,0) ;
      X = J(NGMOVE,1,0) ;
      S =  J(12,1,0) ;
DO K=1 TO NMOVE ;
DO I=1 TO 12 ;
      SUMM(K,1) = SUMM(K,1) + Y(I+K-1,1) #/ 12 ;
END ;                             * I-LOOP ;
END ;                             * K-LOOP ;
DO K=1 TO NGMOVE ;
      YY(K,1)=Y(K+6,1) ;
      X(K,1) = K+6 ;
DO I=1 TO 2 ;
      TC(K,1) = TC(K,1) + SUMM(I+K-1,1) #/ 2 ;
END ;                             * I-LOOP ;
      SI(K,1) = Y(K+6,1) * 100 #/ TC(K,1) ;
END ;                             * K-LOOP ;
DO K=1 TO 12 ;
DO I=1 TO NY1 ;
      S(K,1) = S(K,1) + SI(K+12*(I-1),1) #/ NY1 ;
END ;                             * I-LOOP ;
END ;                             * K-LOOP ;
DO K=1 TO 12 ;
DO I=1 TO NY1 ;
      IR(K+(I-1)*12,1) = SI(K+(I-1)*12,1) #/ S(K,1) ;
END ;                             * I-LOOP ;
END ;                             * K-LOOP ;
      TYPI = SUM(IR) #/ NGMOVE ;
      AVES = SUM(SI) #/ NGMOVE ;
DO I=1 TO NGMOVE ;
      IR(I,1) = IR(I,1) - TYPI(1,1) ;
      SI(I,1) = SI(I,1) - AVES(1,1) ;
END ;                             * I-LOOP ;
MAT = X||YY||TC||SI||IR ;
* PRINT RESULTS ;
NOTE    'Average Seasonal Index' ;
PRINT   S ROWNAME = ROWN ;
```

```
OUTPUT MAT OUT=NEW (RENAME=(COL1=X COL2=Y COL3=TC COL4=SI COL5=I)) ;
PROC    PRINT ;
VAR     X Y TC SI I ;
TITLE1 'X=Month' ;
TITLE2 'Y=DATA, TC=Centered Moving Average' ;
TITLE3 'SI=Specific Seasonal Index and I=Irregular Component' ;

* PLOT RESULTS ;
PROC    GPLOT ;
        PLOT Y*X=1  / HAXIS = 0 TO 132 BY 12 ;
        PLOT TC*X=1 / HAXIS = 0 TO 132 BY 12 ;
        SYMBOL1 I=JOIN L=1 C=BLACK ;
        PLOT SI*X=2 / VREF=0   HAXIS= 0 TO 132 BY 12
                               VAXIS=-100 TO 100 BY 50 ;
        PLOT  I*X=2 / VREF=0   HAXIS= 0 TO 132 BY 12
                               VAXIS= -0.8 TO 0.4 BY 0.4 ;
        SYMBOL2 I=NEEDLE C=BLACK ;
TITLE ;
TITLE2 ;
```

Data: Time Series 2

```
139.5 139.7 138.7 146.2 139.5 136.3 150.1 157.3 157.8 151.1 144.8 139.6
144.3 151.5 154.7 158.3 160.4 163.8 164.1 157.3 151.4 136.5 140.1 158.5
149.3 154.2 149.3 145.0 148.0 143.9 143.9 148.1 152.9 152.4 154.4 150.9
149.7 147.1 145.7 139.7 141.4 143.8 141.2 139.3 150.0 163.2 157.7 181.3
177.6 197.6 226.8 209.5 197.7 168.6 167.6 169.8 154.6 138.2 131.6 139.2
143.7 138.9 168.7 181.3 203.4 218.2 241.5 212.6 195.6 217.2 239.5 223.9
230.4 213.7 220.3 261.8 278.0 269.7 338.4 297.2 310.5 289.8 299.7 333.8
339.7 353.2 361.4 379.8 366.4 357.0 388.1 382.8 369.2 393.0 394.5 427.6
407.0 357.2 367.6 396.6 431.3 459.4 489.5 529.7 558.7 573.0 554.5 625.9
659.1 772.5 810.1 772.4 843.6 773.1 882.3 891.6 997.5 972.3 952.7 1112.5
1052.3 1096.6 1128.6 1060.6 1174.1 1244.5 1423.7 1586.6 1505.1 1659.1
1636.3 1634.2
1511.8 1382.0 1609.8 1704.1 1698.2 1564.1 1634.5 1712.8 1821.5 1844.3
1772.3 1788.8;
```

Output: Time Series 1

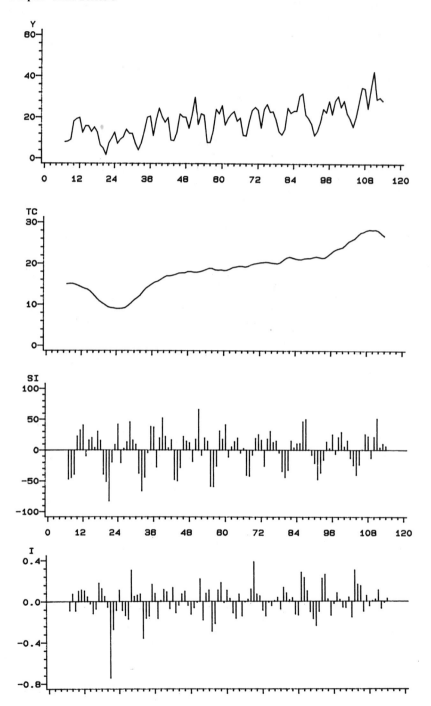

Output: Time Series 2

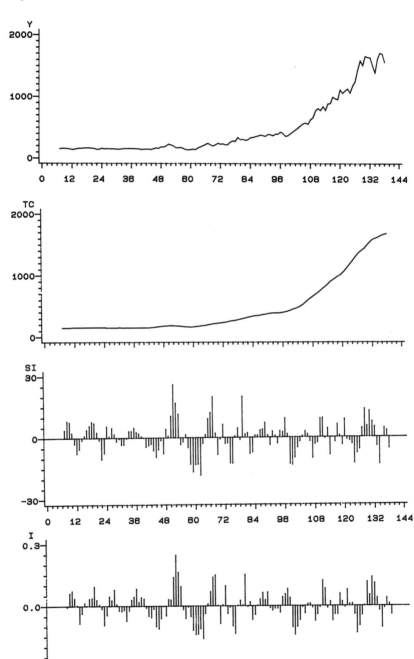

Apart from an increasing general trend the wool export series evidences a clear seasonal pattern. In the wool export figure the irregular variation I for March 1974 appears to be an outlier. Consequently the value of wool exports remained constant from February to March 1974 as opposed to substantial increases during the corresponding months of the other years. The SI and I components of the coal shares series are virtually identical which indicates the absence of a seasonal effect.

2. Representations in the Frequency Domain

Some graphical representations of the different components of a time series as a function of time were considered in the previous section. The graphical methods of this section are used in an attempt to isolate the most important cyclical patterns or periodicities hidden in a time series. Since such cycles are usually identified by their frequencies these graphs represent the time series in the frequency domain.

The two most informative graphs in this respect are the periodogram and the spectral density function. In order to obtain maximum information from these graphs we assume that the time series to be analized is more or less stationary. This implies that the mean and variance of the time series observations as well as the correlations between different observations are constant over differing time segments. The most important source of non-stationarity in a time series is the presence of a trend. The elimination of this trend usually results in an approximately stationary series.

Using the finite Fourier transform a stationary time series can be decomposed into a sum of sine and cosine waves of different amplitudes and wavelengths

$$Y_t = a_0 + \sum_{k=1}^{m} (a_k \cos \omega_k t + b_k \sin \omega_k t), \qquad t = 1, 2, \ldots, n$$

where $m = n/2$ for n even and $m = (n - 1)/2$ for n odd. The term $(a_k \cos \omega_k t + b_k \sin \omega_k t)$ is called the kth harmonic component of the series. The coefficients a_k and b_k are respectively the amplitudes of the waves $a_k \cos \omega_k t$ and $b_k \sin \omega_k t$. The frequency ω_k is defined by

$$\omega_k = 2\pi k/n$$

and the period of the corresponding component by

$$p_k = 2\pi/\omega_k.$$

The smallest and largest frequencies which can be identified from the time series $Y_1, Y_2, \ldots, Y_n$ are therefore $2\pi/n$ and π with corresponding periods n and 2. Graphs in the frequency domain are therefore represented in the interval $(0, \pi)$.

The coefficients $a_0, a_1, \ldots, a_m$ and $b_1, b_2, \ldots, b_m$ are usually calculated by the method of least squares in terms of the observations $Y_1, Y_2, \ldots, Y_n$ and the corresponding frequencies $\omega_1, \omega_2, \ldots, \omega_m$ as described in, for example, Fishman (1969).

$$I_k = \frac{n}{2}(a_k^2 + b_k^2), \qquad k = 1, 2, \ldots m$$

is the contribution of the kth harmonic component to the total corrected sum of squares of the series namely $\sum_{i=1}^{n}(Y_i - \bar{Y})^2$. The graphical representation of I_k against ω_k is known as the periodogram of the time series and gives an indication which harmonic components or cycles are the most important in explaining the variation. Unfortunately the periodogram fluctuates to such an extent that its interpretation is usually difficult. Thus the periodogram is usually smoothed by a weighted moving average to produce a clearer picture of the underlying cycles. This procedure results in an estimate of the so-called spectral density function of the time series. The weight functions are known as spectral windows and a wide variety of these functions have been defined and are listed in Fishman (1969).

The two time series introduced in the previous section will now be analized by means of the SAS PROC SPECTRA program. From the graphs in the time domain it is apparent that both series include a trend while the first one also shows a seasonal component. In order to confirm this and to investigate if other than seasonal harmonic components are present a spectral analysis is worthwhile.

Input: Time Series 1

```
DATA    WOOL ;
INPUT   MONTH WOOLEXP @@ ;
CARDS ;
001 13.4   002 13.1   003 17.9   004 17.0   005 18.2   006 18.3
007  7.9   008  8.3   009  9.1   010 18.1   011 19.2   012 19.9
013 12.5   014 15.8   015 15.8   016 12.9   017 15.1   018 12.6
019  6.3   020  4.8   021  1.6   022  7.4   023  9.9   024 12.7
025  7.1   026  9.2   027 10.4   028 14.1   029 12.0   030 12.1
031  7.2   032  4.0   033  7.3   034 13.3   035 20.0   036 20.6
037 11.1   038 18.7   039 24.4   040 20.3   041 17.5   042 19.8
043  8.8   044  8.6   045 12.6   046 21.6   047 20.2   048 20.0
049 14.6   050 21.0   051 29.7   052 16.5   053 21.7   054 21.1
055  7.7   056  7.6   057 13.6   058 24.0   059 21.7   060 25.8
061 16.3   062 19.6   063 21.6   064 22.9   065 18.3   066 19.8
067 11.3   068 11.1   069 18.1   070 23.6   071 25.0   072 23.5
073 15.0   074 24.0   075 26.4   076 22.6   077 23.0   078 19.0
079 13.3   080 11.6   081 14.4   082 24.7   083 22.2   084 23.3
085 23.2   086 30.6   087 31.8   088 21.4   089 19.6   090 16.9
091 11.3   092 13.5   093 17.9   094 24.3   095 22.5   096 28.1
097 21.6   098 28.3   099 30.5   100 25.2   101 28.3   102 22.1
103 19.4   104 15.6   105 20.5   106 27.5   107 34.7   108 34.1
109 24.5   110 34.0   111 42.5   112 29.0   113 29.8   114 28.2
115 17.8   116 15.7   117 20.9   118 20.9   119 25.3   120 30.0
;
```

```
PROC    REG ;
        MODEL WOOLEXP=MONTH / DW P R ;
        OUTPUT OUT=A P=TREND R=RESIDUAL ;
PROC    PLOT DATA=A ;
        PLOT WOOLEXP*MONTH ='A' TREND*MONTH='T' / OVERLAY
                                                    VPOS=24 HPOS=60 ;
        PLOT RESIDUAL*MONTH ='A' / VPOS=24 HPOS=60 ;
PROC    SPECTRA DATA=WOOL P S OUT=B ;
        VAR WOOLEXP ;
        WEIGHTS 1 2 3 4 3 2 1 ;
PROC    PLOT DATA=B ;
        PLOT (P_01 S_01)*(FREQ PERIOD) / VPOS=24 HPOS=60 ;
PROC    SPECTRA DATA=A P S OUT=C ;
        VAR RESIDUAL ;
        WEIGHTS 1 2 3 4 3 2 1 ;
PROC    PLOT DATA=C ;
        PLOT (P_01 S_01)*(FREQ PERIOD) / VPOS=24 HPOS=60 ;
```

Output: Relevant output for spectral analysis of Time Series 1

Since the observed series is nonstationary a straight line was fitted by means
of the SAS PROC REG program. From the residual plot it appears that the
series of residuals is stationary so that the subsequent analysis was performed
on the residuals.

(i) *Residual plot*

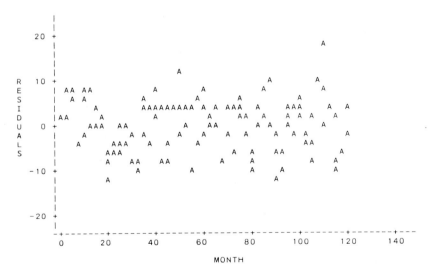

The spectral density function plots were obtained by using a weighted moving
average of the I_k-values. The weights for $\{I_{k-3}, I_{k-2}, I_{k-1}, I_k, I_{k+1}, I_{k+2}, I_{k+3}\}$
that were used are $\{1, 2, 3, 4, 3, 2, 1\}$. The spectral density plots are given in
terms of the frequencies ω_k as well as in terms of the periods p_k. In order to
show that nonstationarity could be detected by a spectral analysis the analysis
was also performed on the originally observed series.

(ii) *Spectral density plots for original data*

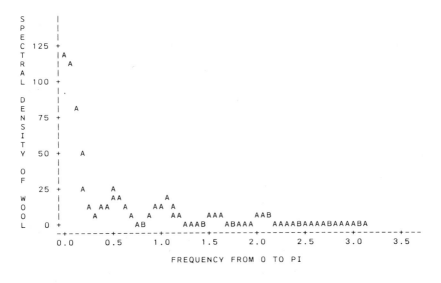

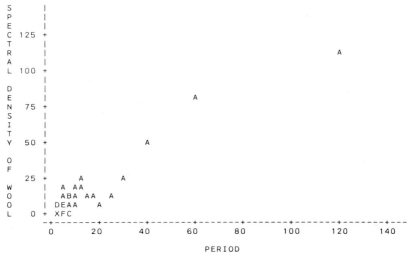

The above two plots show a cycle with a very small frequency (or a large period) which reveals the existence of a trend in the series.

(iii) *Periodogram for residual series*

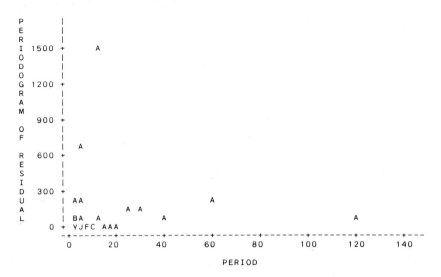

(iv) *Spectral density plots for residual series*

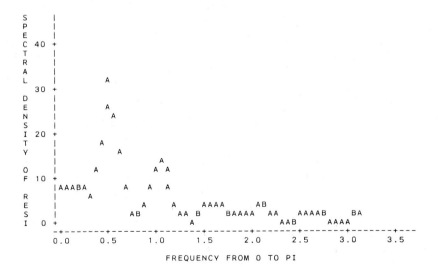

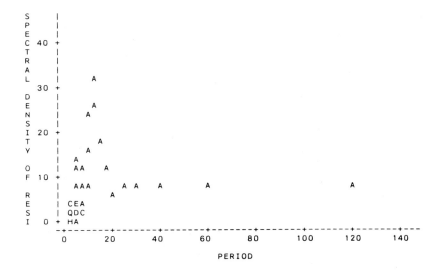

From the two spectral density plots of the residual series the most prominent cycle has an approximate frequency of 0.5 which corresponds to a 12-month period. A less prominent cycle corresponding to a 6-month period is also evident from the above two plots. Conclusions such as the above can also be drawn from the periodogram for the residual series.

Input: Time Series 2

```
DATA    COAL ;
INPUT   MONTH COALIND @@ ;
        IF MONTH < 61 THEN TRENCOAL = 154.2 ;
        IF MONTH >= 61 THEN TRENCOAL=25.957*((1.03079)**MONTH) ;
        DETCOAL=(COALIND/TRENCOAL)*100 ;
CARDS ;
 001 140  002 140  003 139  004 146  005 140  006 136
 007 150  008 157  009 158  010 151  011 145  012 140
 013 144  014 152  015 155  016 158  017 160  018 164
 019 164  020 157  021 151  022 137  023 140  024 159
 025 149  026 154  027 149  028 145  029 148  030 144
 031 144  032 148  033 153  034 152  035 154  036 151
 037 150  038 147  039 146  040 140  041 141  042 144
 043 141  044 139  045 150  046 163  047 158  048 181
 049 178  050 198  051 227  052 210  053 198  054 169
 055 168  056 170  057 155  058 138  059 132  060 139
 061 144  062 139  063 169  064 181  065 203  066 218
 067 242  068 213  069 196  070 217  071 240  072 224
 073 230  074 214  075 220  076 262  077 278  078 270
 079 338  080 297  081 311  082 290  083 300  084 334
 085 340  086 353  087 361  088 380  089 366  090 357
 091 388  092 383  093 369  094 393  095 395  096 428
 097 407  098 357  099 368  100 397  101 431  102 459
 103 490  104 530  105 559  106 573  107 555  108 626
 109 659  110 773  111 810  112 772  113 844  114 773
 115 882  116 892  117 998  118 972  119 953  120 1113
 121 1052 122 1097 123 1129 124 1061 125 1174 126 1245
 127 1424 128 1587 129 1505 130 1659 131 1636 132 1634
 133 1512 134 1382 135 1610 136 1704 137 1698 138 1564
 139 1635 140 1713 141 1822 142 1844 143 1772 144 1789
;
PROC    PLOT ;
        PLOT COALIND*MONTH='A' TRENCOAL*MONTH='T'/
                          OVERLAY VPOS=24 HPOS=60 ;
        PLOT DETCOAL*MONTH / VPOS=24 HPOS=60 ;
PROC    SPECTRA P S OUT=A ;
        VAR DETCOAL ;
        WEIGHTS 1 2 3 4 3 2 1 ;
PROC    PLOT DATA=A ;
        PLOT (P_01 S_01)*(FREQ PERIOD) / VPOS=24 HPOS=60 ;
```

The TC-plot in Section 1 for the index of coal shares suggests a constant trend for 1970–1974. The years 1975–1981 show an exponential increase in market prices. The trend fitted was therefore

$$\hat{Y}_t = \begin{cases} 154.2 & t \le 60 \\ 25.957(1.03079)^t, & t > 60. \end{cases}$$

The trend was eliminated by means of the transformation $(Y_t/\hat{Y}_t) \times 100\%$ which we shall call the detrended series.

Output: Relevant output for the spectral analysis for Time Series 2

A similar analysis as that for Time Series 1 applied to the output given below shows that the coal shares index contains no seasonal effect. The only possible cycle is one of at least three years which can be termed the so-called cyclical movement and which can be related to the economic activity during this period.

(i) *Plot of detrended index of coal shares*

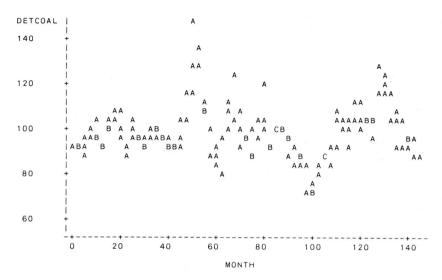

(ii) *Spectral density plots of detrended series*

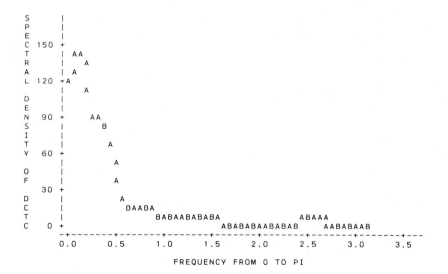

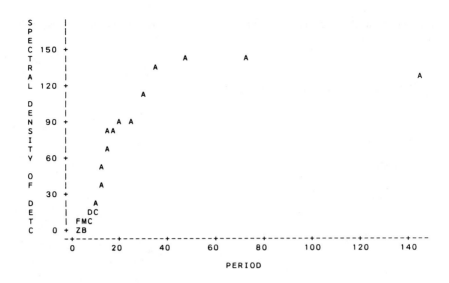

PERIOD

CHAPTER 11

Further Useful Graphics

1. Graphics for the Two-Sample Problem

It is often important to compare two univariate samples in respect of locality and variation. The *P-P* (Probability-Probability) plot is a simple and informative method for drawing such a comparison. By using a univariate or bivariate scaling of multivariate data however, a *P-P* plot can also be used for comparing two multivariate samples with each other. The univariate and multivariate cases will now be discussed.

A. *P-P* Plot for Two Univariate Samples

Suppose $X_1, X_2, \ldots, X_m$ and $Y_1, Y_2, \ldots Y_n$ are two independent samples. Furthermore, let F and G represent the two empirical cumulative distribution functions of the respective samples, in other words

$$\hat{F}(x) = i/m$$

if i of the X values are smaller than or equal to x;

$$\hat{G}(y) = j/n$$

if j of the Y values are smaller than or equal to y.

A *P-P* plot of the two samples is obtained by plotting the points $(\hat{G}(z), \hat{F}(z))$ where z represents the different observations in the joint sample of size $m + n$. The name *P-P* plot issues from the fact that this representation shows the empirical probabilities for every z value of finding an X or Y value in the respective samples which is smaller than or equal to z.

In order to explain the construction and interpretation of a *P-P* plot, we

consider the scores of the 16 boys and 12 girls in the Junior Aptitude Test
(Reasoning) (see Table 1.1—X_{19}).

Boys ($m = 16$)

38, 34, 45, 42, 28, 23, 42, 24, 27, 30, 31, 31, 34, 34, 40, 41

Girls ($n = 12$)

39, 42, 35, 38, 33, 27, 33, 37, 31, 33, 38, 39

Ordered Joint Sample (z-Values)

23(b), 24(b), 27(g), 27(b), 28(b), 30(b), 31(b), 31(b), 31(g), 33(g),

33(g), 33(g), 34(b), 34(b), 34(b), 35(g), 37(g), 38(g), 38(b), 38(g),

39(g), 39(g), 40(b), 41(b), 42(g), 42(b), 42(b), 45(b).

The symbols b and g between brackets respectively indicate whether the
observation derives from the boys' or girls' sample. For equal values in the
two samples, the order in the joint sample is determined randomly. The step
function in Figure 11.1 was obtained by plotting the points $(\hat{G}(z), \hat{F}(z))$, where
z assumes the values 23, 24, ..., 45. What this amounts to, is that vertical lines
of length 1/16 and horizontal lines of length 1/12 are respectively drawn in
the same order as the occurrence of the observations in respect of the boys
and girls in the ordered joint sample. The *P-P* plot is completed by drawing
the 45° line $\hat{F}(z) = \hat{G}(z)$ which acts as the reference line.

Two samples derived from identical populations will have a *P-P* function
that differs little from the 45° line as in Figure (a) below. Locality differences

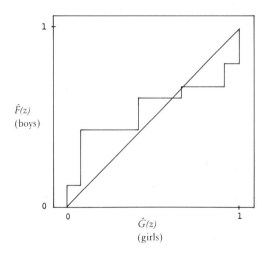

Figure 11.1. *P-P* plot of the Junior Aptitute Test scores for boys and girls.

between the two samples are characterized by *P-P* functions mainly lying either far above or far below the 45° line [see Figure (b)]. However, a *P-P* function which crosses the 45° line from the one side to the other [see Figure (c)], indicates a difference in variation between the two samples. Figure 11.1 shows a possible difference in variation rather than in locality between the two samples. It is obvious that the boys' test scores vary more than those of the girls.

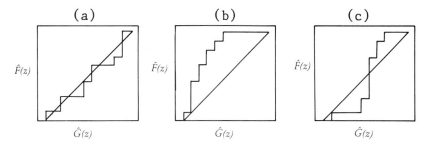

B. *P-P* Plot for Two *p*-Variate Samples

Consider the following independent data matrices **X** and **Y** consisting of observations on the same set of variables:

$$
\mathbf{X} = \begin{bmatrix} X_{11} & X_{12} & \cdots & X_{1p} \\ X_{21} & X_{22} & \cdots & X_{2p} \\ \vdots & \vdots & & \vdots \\ X_{m1} & X_{m2} & \cdots & X_{mp} \end{bmatrix}, \qquad \mathbf{Y} = \begin{bmatrix} Y_{11} & Y_{12} & \cdots & Y_{1p} \\ Y_{21} & Y_{22} & \cdots & Y_{2p} \\ \vdots & \vdots & & \vdots \\ Y_{n1} & Y_{n2} & \cdots & Y_{np} \end{bmatrix}.
$$

The joint sample of the $(m + n)$ *p*-variate observations is then represented by

$$
\begin{pmatrix} \mathbf{X} \\ \mathbf{Y} \end{pmatrix}.
$$

In Chapter 6 various methods were discussed according to which a *p*-variate data set can be represented by means of a scatterplot in two dimensions or even as a one-dimensional ordering.

If a one-dimensional scaling explains a large percentage of the variation in the joint sample, a *P-P* plot as explained in A can be used for comparing the two multivariate samples (reduced now to two univariate samples) with each other. Figure 11.2 gives the *P-P* plot for the 16 boys and 12 girls in respect of the 18 ability test variables and is based on the one-dimensional non-metric scaling of Chapter 6, Section 6. The plot does not show any clearly observable differences in respect of locality and variation between the two data sets.

In the case of a two-dimensional scaling it is also possible to use the *P-P* plot in order to detect differences in spread and locality between two data matrices. The method is described by Friedman and Rafsky (1979, 1981) and

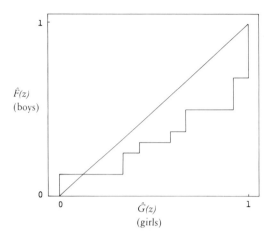

Figure 11.2. *P-P* plot of one-dimensional scaling of 18 ability tests for boys and girls.

entails using ranks based on a minimal spanning tree in order to obtain a one-dimensional ordering of the two-dimensional scaling points. Suppose that the eccentricity of a node in a *MST* is the number of edges in a path with greatest length beginning with that node. The end point of this longest path is called the antipode of the node and the path between a node with largest eccentricity and its antipode is called a diameter. Finally, define the center node as the node with the smallest eccentricity.

In order to detect a possible difference in locality between two data matrices **X** and **Y** it is necessary to assign locality ranks to the nodes of the minimal spanning tree of the combined sample. The ranks are assigned as follows:

(i) Root the tree at one end of the diameter.
(ii) Beginning at the root, follow the diameter and assign ranks 1 to n to the nodes corresponding to the order in which they are visited. Subtrees (branches) of the tree are visited along the way and ties are resolved by first visiting branches which are closer in Euclidean distance to the root.

Figure 11.3 gives the minimal spanning tree based on the two-dimensional non-metric scaling of the Ability Test Data set of Table 1.1 (see Section 6, Chapter 6). Figure 11.4 shows the locality rankings of the 28 pupils where each point is classified as boy or girl. The root is taken as the two-dimensional representation of the scores of Pupil Number 16 and therefore has rank 1. The antipode is Pupil Number 24. The *P-P* plot of these rankings appears in Figure 11.5. No noteworthy locality differences between the data sets of the girls and boys is evident from this plot.

In order to investigate the possibility that data sets **X** and **Y** differ in spread, radial ranks must be assigned to the nodes of the minimal spanning tree given in Figure 11.3. The procedure entails the following:

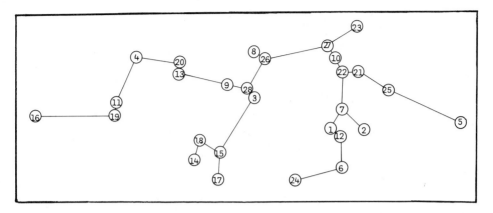

Figure 11.3. Minimal spanning tree for Ability Test Data based on a two-dimensional non-metric scaling.

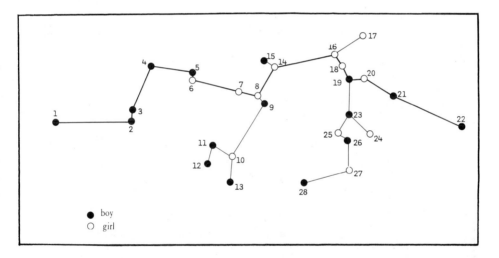

Figure 11.4. Locality rankings of Ability Test Data.

(i) Root the MST on the center node (the node with the smallest eccentricity) and assign rank 1 to this node.

(ii) Ranks are now assigned to each node in the order of the number of edges between the node and the center. Ties are again resolved by giving the node nearest (in an Euclidean sense) to the root the smallest rank.

Since the center node lies near the geometric center of the multivariate sample, radial ranks measure the nodes' deviations from the midpoint. Due to the way in which radial ranks are obtained, a *P-P* plot based on these ranks seemingly

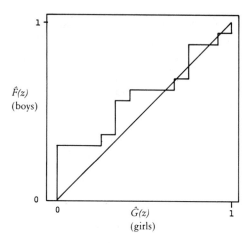

Figure 11.5. *P-P* plot of locality rankings of Ability Test Data for boys and girls.

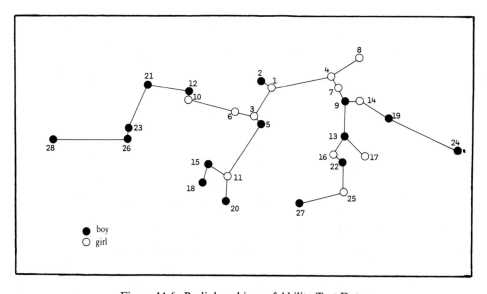

Figure 11.6. Radial rankings of Ability Test Data.

shows up a difference in spread between two samples as a locality difference. Figure 11.6 shows the MST with radial ranks and Figure 11.7 the corresponding radial *P-P* plot. This plot shows an obvious difference in spread between the Ability Test Data sets for boys and girls. From these two figures it appears that the boys' scores have a larger variation than the girls' scores.

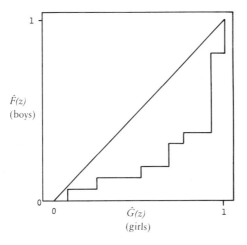

Figure 11.7. Radial *P-P* plot of radial rankings of Ability Test Data for boys and girls.

2. Graphical Techniques in Analysis of Variance

The aim of analysis of variance is to determine whether the means of k populations differ from one another. The different populations are usually associated with different treatments which are carried out independently of one another on homogeneous experimental units. It is also assumed that the n_1 observations of the first treatment can be regarded as a random sample from a $n(\mu_1, \sigma^2)$ population, the n_2 observations of the second treatment as a random sample from a $n(\mu_2, \sigma^2)$ population, ..., and the n_k observations of the kth treatment as a random sample from a $n(\mu_k, \sigma^2)$ population.

In this section three graphical techniques are discussed on the basis of a completely randomized design with a view to

○ assessing the assumption of normality, and
○ determining possible differences between the treatment means.

A. Assessing the Assumption of Normality

The assumption of normality can be assessed by means of a normal probability plot for each of the k groups of observations and evaluating the linearity of each graphical representation. An easier way however, is to standardize each of the k sets of observations and then to draw a normal probability plot of the joint sample of $n_1 + n_2 + \cdots + n_k$ observations. If the assumption of normality does apply for all k samples this plot should indicate a linear trend.

If Y_{ij} represents the ith observation of the jth treatment, standardization involves transforming Y_{ij} to Z_{ij}, where

$$Z_{ij} = (Y_{ij} - \bar{Y}_j)/s_j; \qquad j = 1, 2, \ldots, k; \quad i = 1, 2, \ldots, n_j,$$

with $\bar{Y}_j$ and s_j respectively the arithmetic mean and standard deviation of the jth set of observations.

In order to illustrate the above mentioned technique we consider random samples of size 10 each from the four groups of pupils in the joint data set of 2800 pupils. Here the variable of concern is one of the ability tests, namely X_{18} (Word Analogies—Education Level III). Table 11.1 gives the four groups of observations (Y), the group means, standard deviations as well as the standardized value (Z) of each observation. Table 11.2 gives the cumulative frequency table of the 40 standardized values. The normal probability plot in

Table 11.1. Scores of four groups of pupils in the Word Analogies Test (Educational Level III)

	Language group A, boys		Language group A, girls		Language group B, boys		Language group B, girls	
	Y_{i1}	Z_{i1}	Y_{i2}	Z_{i2}	Y_{i3}	Z_{i3}	Y_{i4}	Z_{i4}
	18	0.258	16	−0.108	14	−0.338	21	1.289
	19	0.627	18	0.609	15	−0.056	19	0.660
	19	0.627	13	−1.183	14	−0.338	17	0.031
	13	−1.587	18	0.609	10	−1.465	19	0.660
	19	0.627	12	−1.541	15	−0.056	16	−0.283
	15	−0.849	13	−1.183	16	0.225	11	−1.855
	13	−1.587	20	1.326	10	−1.465	18	0.346
	20	0.996	18	0.609	21	1.634	20	0.975
	17	−0.110	16	−0.108	19	1.070	13	−1.226
	20	0.996	19	0.968	18	0.789	15	−0.597
$\bar{Y}_j$:	17.30		16.30		15.20		16.90	
s_j:	2.71		2.79		3.55		3.18	

Table 11.2. Cumulative frequency distribution of Z-values in Table 11.1

Class limit	F	$F/(n + 1)$
−1.3	6	0.146
−0.7	11	0.268
−0.1	17	0.415
0.5	23	0.561
1.1	37	0.902
1.7	40	0.976

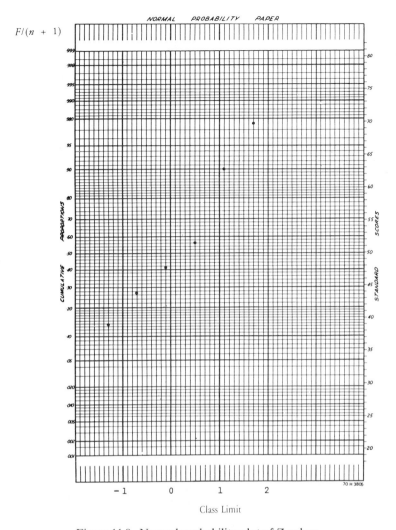

Figure 11.8. Normal probability plot of Z-values.

Figure 11.8 indicates a relatively linear pattern, which means that the observations probably derive from normal populations.

B. Determining Possible Differences in the Treatment Means

In order to detect possible differences in the mean effect of treatments in the case where $n_1 = n_2 = \cdots = n_k = n$, a normal probability plot is made of the ordered values of $\bar{Y}_1, \bar{Y}_2, \ldots, \bar{Y}_k$. If no treatment differences occur, this plot should take the form of a straight line since $\bar{Y}_1, \bar{Y}_2, \ldots, \bar{Y}_k$, in the case of large

Table 11.3. Ordered group means and
rankits for X_{18}

Mean	15.2	16.3	16.9	17.3
Rankit	−1.029	−0.297	0.297	1.029

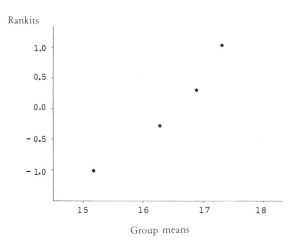

Figure 11.9. Plot of rankits against ordered group means.

n, have approximately the same normal distribution. Deviations from linearity point to possible differences between treatment means.

As an example consider again the observations on X_{18} of the four groups of pupils at Educational Level III. Table 11.3 shows the ordered group means as well as the rankits (see Chapter 3). Figure 11.9 is the graphical representation of the information contained in Table 11.3.

Although the four points do not exhibit an exact linear relation, the figure nevertheless shows that no large differences in group means are present.

C. Pairwise Comparisons

Pairwise comparison of treatment means $\bar{Y}_1, \bar{Y}_2, \ldots, \bar{Y}_k$ is based on a set of intervals of the form

$$(\bar{Y}_m - \bar{Y}_j) \pm l_{mj}, \qquad m < j$$

where l_{mj} is the length of the interval and is a function of the number of observations $n_1, n_2, \ldots, n_k$. If the interval excludes 0 it is concluded that $\bar{Y}_m - \bar{Y}_j$, at a prescribed level of significance, differ significantly from zero.

In the case where $n_1 = n_2 = \cdots = n_k = n$, the interval length l_{mj} is constant for all pairwise comparisons and is indicated by l. Pairwise comparison now

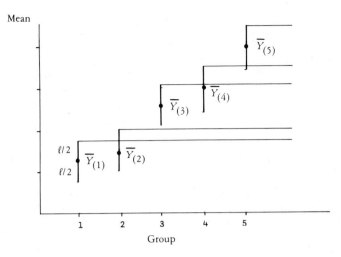

Figure 11.10. Ordered sample means and intervals for pairwise comparisons.

occurs graphically as follows [Hochberg, Weiss and Hart (1982)]:

Arrange the treatment means from small to large and let the ordered values be $\bar{Y}_{(1)}, \bar{Y}_{(2)}, \ldots, \bar{Y}_{(k)}$. The intervals, each of length l, are subsequently set perpendicularly around the plotted points in such a way that the centre of each vertical line corresponds to a plotted point. Next, draw horizontal lines to the right from the top of each interval.

Figure 11.10 is an example of the pairwise comparison of five treatment means. Any pair of treatments of which the intervals are not joined by a common horizontal line, differ significantly.

From Figure 11.10 it subsequently follows that $\bar{Y}_{(1)}$ and $\bar{Y}_{(2)}$ do not differ from each other, but do differ significantly from $\bar{Y}_{(3)}$, $\bar{Y}_{(4)}$ and $\bar{Y}_{(5)}$. $\bar{Y}_{(3)}$ and $\bar{Y}_{(5)}$ also differ significantly but not $\bar{Y}_{(4)}$ and $\bar{Y}_{(5)}$.

Consider now Scheffe's procedure of pairwise comparison. Assume $N = n_1 + n_2 + \cdots + n_k$. The length of Scheffe's interval is calculated as follows:

$$l_{mj}^2 = \left(\frac{1}{n_m} + \frac{1}{n_j}\right)\left(\frac{k-1}{N-k}\right)S^2 F_{k-1,N-k;0.05}.$$

S^2 is the error sum of squares and is calculated as

$$S^2 = \sum_{j=1}^{k}\sum_{i=1}^{n_j} Y_{ij}^2 - \sum_{j=1}^{k} n_j \bar{Y}_j^2.$$

$F_{k-1,N-k;0.05}$ is the 5% upper critical value of the F distribution with $k-1$ and $N-k$ degrees of freedom.

If $n_1 = n_2 = \cdots = n_k = n$, the following interval of constant length is obtained:

$$l^2 = \frac{2}{n}\left(\frac{k-1}{N-k}\right)S^2 F_{k-1,N-k;0.05}.$$

A conservative graphical method (which usually gives fewer significant differences than are actually present) in the case of an unequal number of observations per treatment, is obtained by calculating an interval length l' where

$$(l')^2 = \left(\frac{1}{n_{(1)}} + \frac{1}{n_{(2)}}\right)\left(\frac{k-1}{N-k}\right)S^2 F_{k-1,N-k;0.05}.$$

In the above expression $n_{(1)}$ and $n_{(2)}$ are the smallest and second smallest sample sizes.

Consider the data of Table 11.1. $\bar{Y}_{(1)}$, $\bar{Y}_{(2)}$, $\bar{Y}_{(3)}$ and $\bar{Y}_{(4)}$, respectively, indicate the means in respect of boys from Language Group B, girls from Language Group A, girls from Language Group B and boys from Language Group A. Now,

$$l^2 = \frac{8}{40}\left(\frac{3}{36}\right)S^2 F_{3,36;0.05},$$

where

$$S^2 = \sum_{j=1}^{4}\sum_{i=1}^{10} Y_{ij}^2 - 10\sum_{j=1}^{4}\bar{Y}_j^2 = 11{,}157 - 10{,}816.3 = 340.7.$$

Consequently $l = 4.06$ for $F_{3,36;0.05} = 2.9$.

Figure 11.11 is a graphical representation of the pairwise comparison of the four groups of pupils. This plot further confirms that there are no signi-

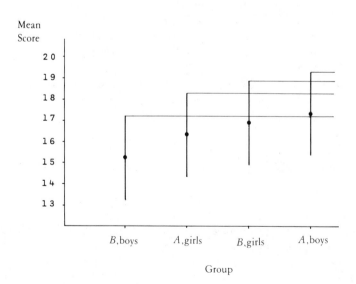

Figure 11.11. Pairwise comparison of the mean score of the four groups of pupils in X_{18}.

ficant differences between the average achievements of the four groups for variable X_{18}.

3. Four-Fold Circular Display of 2 × 2 Contingency Tables

A 2 × 2 table of observed counts or observed percentages is usually displayed graphically by means of bar charts or pie charts (see Chapter 2). An alternative way of graphical representation was considered by Fienberg (1975). According to this method, cell percentages are presented by quarter circles in such a way that the areas are proportional to these percentages. An advantage of this type of representation is that the interaction structure of the table is depicted graphically.

Consider the following table of percentages:

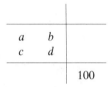

Given that the ratios of areas $\frac{1}{4}\pi r_a^2 : \frac{1}{4}\pi r_b^2 : \frac{1}{4}\pi r_c^2 : \frac{1}{4}\pi r_d^2$ are equal to the ratios of percentages $a : b : c : d$, it follows that the areas of the quarter circles will be proportional to the percentages if we chose as radii r_a, r_b, r_c and r_d multiplied by any common positive number.

To illustrate this circular display, consider the following table of language characteristics [Fienberg (1975)]:

Command of Languages

		Bilingual	Unilingual	Totals
Language group affiliation	English	6.4	71.9	78.3
	French	12.4	9.3	21.7
	Totals	18.8	81.2	100

The areas of the quarter circles will be proportional to the percentages if the radii are chosen as $\sqrt{6.4}$, $\sqrt{71.9}$, $\sqrt{12.4}$ and $\sqrt{9.3}$, that is 2.53, 8.48, 3.52 and 3.05. To obtain a suitable scale for the graphical representation each of these values were multiplied by a common constant.

Comparisons between adjacent quarter circles in terms of the radii are carried out visually on a square-root scale.

From the graphical representation in Figure 11.12 the interaction between language group and command of languages is clearly evident: More than half

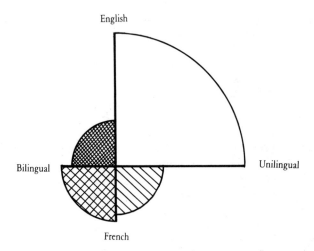

Figure 11.12. Four-fold circular display of 2 × 2 table of percentages.

of the French language group is bilingual whereas only a small percentage of the English language group is bilingual.

The same graphical technique may also be applied to other forms of the 2 × 2 table, e.g., a standardized table in which percentages sum to 100 across rows and across columns. This table preserves the interaction structure present in the original table and the cell entries are obtained by means of iterative proportional fitting [Fienberg (1978)].

The standardized table of language characteristics is as follows:

| | | Command of languages | | |
		Bilingual	Unilingual	Totals
Language group affiliation	English	20.5	79.5	100
	French	79.5	20.5	100
	Totals	100	100	200

Figure 11.13 is a graphical representation of the standardized table. The interaction structure is clearly evident and the display exhibits a symmetric form corresponding to the diagonal and off-diagonal entries of the table.

Extension to 2 × 2 × k Tables

Table 11.4 gives a set of five 2 × 2 tables and represents a subset of a more comprehensive data set consisting of information on two symptoms (breathlessness and wheeze) and age (Fienberg (1978), p. 57). On the right hand side of each table is its corresponding standardized table.

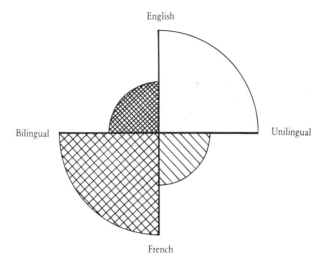

Figure 11.13. Four-fold circular display for 2 × 2 standardized table of percentages.

Table 11.4. Coal miners classified by age (in 5-year groupings), breathlessness (B) and wheeze (W)

Observed table				Standardized table	
			Age 25–29		
	W	No W		W	No W
B	23	9	B	86.4	13.6
No B	105	1654	No B	13.6	86.4
			Age 30–34		
	W	No W		W	No W
B	54	19	B	84.5	15.5
No B	177	1863	No B	15.5	84.5
			Age 35–39		
	W	No W		W	No W
B	121	48	B	82.8	17.2
No B	257	2357	No B	17.2	82.8
			Age 40–44		
	W	No W		W	No W
B	169	54	B	81.9	18.1
No B	273	1778	No B	18.1	81.9
			Age 45–49		
	W	No W		W	No W
B	269	88	B	80.1	19.9
No B	324	1712	No B	19.9	80.1

The method for obtaining the standardized tables will now be illustrated by means of the following SAS PROC MATRIX program and by using the observed 2 × 2 table for the 25–29 year age group.

Input: SAS PROC MATRIX program for the standardization of a contingency table

```
PROC    MATRIX ;
        COLN = 'Error' 'Maxdev'  'No. iter' ;
        DIM = 2 2 ;                    * Dimension of table ;
        INITAB = 23 9 / 105 1654 ;    * Observed contingency table ;
        MOD = 0.01 30 ;               * Max deviation , max iterations ;
        TABLE = 50 50 / 50   50 ;     * Table providing marginal totals ;
        CONFIG = 1 2 ;                * Specifies which marginals to fit ;
CALL    IPF(FIT,STATUS,DIM,TABLE,CONFIG,INITAB,MOD) ;
NOTE    'Convergence information' ;
PRINT   STATUS COLNAME = COLN ;
NOTE    'Standardized contingency table' ;
PRINT   FIT ;
```

Output:

Convergence information

STATUS	Error	Maxdev	No. iter
ROW 1	0	0.00822605	15

Standardized contingency table

FIT	COL 1	COL 2
ROW 1	86.387	13.613
ROW 2	13.6173	86.3827

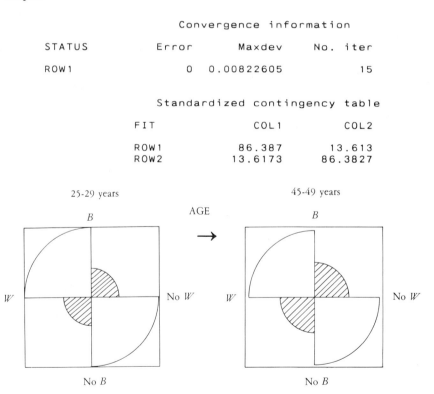

Figure 11.14. Summarization of age dependent interaction between breathlessness and wheeze.

A study of the cell entries of the standardized 2 × 2 tables given in Table 11.4 reveals that the diagonal entries decrease with age and vice versa for the non-diagonal entries. Circular displays of the 2 × 2 standardized contingency tables for age groups 25–29 and 45–49 are given in Figure 11.14. The pattern of change in interaction as age increases is clearly evident and is also summarized in this display.

This general trend holds for the original complete data set, but in this case the results are not as clear-cut as for the subdata set.

References

Anderberg, M.R. (1973). *Cluster analysis for applications*. New York: Academic Press.

Anderson, E. (1960). A semigraphical method for the analysis of complex problems. *Technometrics* 2: 387–391.

Andrews, D.F. (1972). Plots of high-dimensional data. *Biometrics* 28: 125–136.

Anscombe, F.J. (1973). Graphs in statistical analysis. *The Amer. Statist.* 27: 17–22.

Barnett, V. (1975). Probability plotting methods and order statistics. *Applied Statistics* 24: 95–108.

Becker, R.A. and Chambers, J.M. (1977). GR-2: A system of graphical subroutines for data analysis. In: *Computer science and statistics*: 10*th annual symposium on the interface*. National Bureau of Standards. Special Publication 503: 409–415.

Belsley, D.A., Kuh, E. and Welsch, R.E. (1980). *Regression diagnostics*. New York: John Wiley.

Canter, D. (ed.) (1985). *Facet theory approaches to social research*. New York: Springer-Verlag.

Carroll, J.D. and Chang, J.J. (1970). Analysis of individual differences in multidimensional scaling via an, *n*-way generalization of 'Eckart-Young' decomposition. *Psychometrika* 35, 283–319.

Chambers, J.M., Cleveland, W.S., Kleiner, B. and Tukey, P.A. (1983). *Graphical methods for data analysis*. Belmont: Wadsworth.

Chernoff, H. (1973). Using faces to represent points in K-dimensional space graphically. *J. Amer. Statist. Assoc.* 68: 361–368.

Chernoff, H. and Rizvi, M.H. (1975). Effect on classification error or random permutations of features in representing multivariate data by faces. *J. Amer. Statist. Assoc.* 70: 548–554.

Cleveland, W.S. and McGill, R. (1984a). Graphical Perception: Theory, experimentation and application to the development of graphical methods. *J. Amer. Statist. Assoc.* 79: 531–554.

Cleveland, W.S. and McGill, R. (1984b). The many faces of a scatterplot. *J. Amer. Statist. Assoc.* 79: 807–822.

Cleveland, W.S. and Terpenning, I.J. (1982). Graphical methods for seasonal adjustment. *J. Amer. Statist. Assoc.* 77: 52–62.

Cohen, A. (1980). On the graphical display of the significant components in two-way contingency tables. *Comm. Statist.—Theor. Meth.* A9: 1025–1041.

Cook, R.D. and Weisberg, S. (1982). *Residuals and influence in regression.* New York: Chapman and Hall.

Cudeck, R.C. (1982). The evaluation of diagnostic models. *S.A.J. of Psychology*, 12: 61–64.

Daniel, C. and Wood, F.S. (1971). *Fitting equations to data.* New York: John Wiley.

Dixon, W.J. et al. (1983). *Biomedical Computer Programs. P-Series.* Berkeley: University of California Press.

Draper, N. and Smith, H. (1981). *Applied regression analysis* (2nd ed.). New York: John Wiley.

du Toit, S.H.C. et al. (1982). *Statistical techniques for the analysis of qualitative and quantitative multivariate data.* HSRC-seminar PR79. Pretoria: HSRC.

Edwards, W.F. and Cavalli-Sforza, L.L. (1965). A method for cluster analysis. *Biometrics* 21: 362–375.

Everitt, B. (1974). *Cluster analysis.* London: Heineman Educational.

Flury, B. and Riedwyl, H. (1981). Graphical representation of multivariate data by means of assymetrical faces. *J. Amer. Statist. Assoc.* 76: 757–765.

Fienberg, S.E. (1975). Perspective Canada as a social report. *Social Indicators Research* 2: 151–174.

Fienberg, S.E. (1978). *The analysis of cross-classified categorical data.* Cambridge (Massachusetts): MIT Press.

Fienberg, S.E. (1979). Graphical Methods in Statistics. *The Amer. Statist.* 33: 165–178.

Fishman G.S. (1969). *Spectral methods in econometrics.* Cambridge (Massachusets): Harvard University Press.

Friedman, J.H. and Rafsky, L. (1979). Multivariate generalizations of the Wald-Wolfowitz and Smirnov two-sample tests. *Ann. Statist.* 7: 697–717.

Friedman, J.H. and Rafsky, L. (1981). Graphics for the multivariate two-sample problem. *J. Amer. Statist. Assoc.* 76: 277–295.

Gabriel, K.R. (1971). The biplot graphic display of matrices with application to principal component analysis. *Biometrika* 58: 453–467.

Gabriel, K.R. (1980). Biplot. In: Johnson, N.L. and Kotz, S. (eds.) *Encyclopedia of Statistical Sciences, Vol. 1.* New York: John Wiley.

Gabriel, K.R. (1981). In: Barnett, V. (ed.). *Interpreting multivariate data.* New York: John Wiley.

Galpin, J.S. and Hawkins, D.M. (1984). The use of recursive residuals in checking model fit in linear regression. *The Amer. Statist.* 38: 94–105.

Gnanadesikan, R. (1977). *Statistical data analysis of multivariate observations.* New York: John Wiley.

Gower, J.C. (1966). Some distance properties of latent root and vector methods used in multivariate analysis. *Biometrika* 53: 325–388.

Gower, J.C. and Ross, G.J.S. (1969). Minimum spanning trees and single linkage cluster analysis. *Applied Statistics* 18: 54–64.

Gower, J.C. (1971). A general coefficient of similarity and some of its properties. *Biometrics* 27: 857–872.

Gratch, N. (ed.) (1973). *25 years of social research in Israel.* Jerusalem: Jerusalem Academic Press.

Greenacre, M.J. (1984). *Theory and applications of correspondence analysis.* London: Academic Press.

Greenacre, M.J. and Underhill, L.G. (1982). Scaling a data matrix in a low-dimensional Euclidean space. In: Hawkins, D.M. (ed.). *Topics in applied multivariate analysis.* Cambridge: Cambridge University Press.

Gunst, R.F. and Mason, R.L. (1980). *Regression analysis and its application.* New York: Marcel Dekker.

Guttmann, L. (1968). A general non-metric technique for finding the smallest coordinate space for a configuration of points. *Psychometrika* 33: 469–506.

Guttman, L. (1977). What is not what in statistics. *The Statistician* 26: 81–107.

Guttman, L. (1980). Recent structural laws of human behavior. *The Bulletin of the Institute of Communications Research* (Keio University) 114: 1–12.

Guttman, L. (1982). What is not what in theory construction. In: Hanser, R.M., Mechanic, D. and Hauer, A. (ed.) *Social structure and behavior*. New York: Academic Press.

Harman, H.H. (1976). *Modern factor analysis* (3rd ed.) Chicago: University of Chicago Press.

Hartigan, J.A. (1975). Printer graphics for clustering. *J. of Statist. Computation and Simulation*. 4: 187–213.

Hawkins, D.M., Muller, M.W. and Ten Krooden, J.A. (1982). Cluster analysis. In: Hawkins, D.M. (ed.) *Topics in applied multivariate analysis*. Cambridge: Cambridge University Press.

Hawkins, D.M. (ed.) (1982). *Topics in applied multivariate analysis*. Cambridge: Cambridge University Press.

Healy, C.C. and Mourton, D.L. (1983). Deviatives of the SDS: Potential Clinical and Evaluative Uses. *Journal of Vocational Behaviour* 23: 318–328.

Herman, R. and Montroll, E.W. (1972). A manner of characterizing the development of countries. *Proc. Nat. Acad. Sci. USA* 69: 3019–3023.

Heymann, C. (1981). XAID-An extended automatic interaction detector. *CSIR Special Report*: SWISK 28. Pretoria: CSIR.

Hoaglin, D.C. (1980). A Poissonness plot. *The Amer. Statist.* 34: 146–149.

Hoaglin, D.C., Mosteller, F. and Tukey, J.W. (1982). *Understanding robust and exploratory data analysis*. New York: John Wiley.

Hochberg, Y., Weiss, G. and Hart, S. (1982). On graphical procedures for multiple comparisons. *J. Amer. Statist. Assoc.* 77: 767–772.

Holland, J.L. (1985). *Making Vocational Choices. A theory of vocational personalities and work environments* (2nd ed.). Englewood Cliffs: Prentice Hall.

Holland, P.W. and Leinardt, S. (1981). An exponential family of probability distributions for directed graphs. *J. Amer. Statist. Assoc.* 76: 33–50.

Hotelling, H. (1947). Multivariate quality control illustrated by the air testing of sample bombsights. In: *Selected techniques of statistical analysis* (Eisenhart et al. eds.) New York: McGraw-Hill.

Jackson, J.E. (1959). Quality control methods for several variables. *Technometrics*. 1: 359–377.

Johnson, N.L. and Kotz, S. (1969). *Discrete distributions*. Boston: Houghton Mifflin Company.

Kass, G.V. (1980). An exploratory technique for investigating large quantities of categorical data. *Applied Statistics* 29: 119–127.

Kleiner, B. and Hartigan, J.A. (1981). Representing points in many dimensions by trees and castles. *J. Amer. Statist. Assoc.* 76: 260–269.

Kruskal, J.B. (1956). On the shortest spanning subtree of a graph, and the trading salesman problem. *Proc. Amer. Math. Soc.* 7: 48–50.

Kruskal, J.B. (1964a). Multidimensional scaling by optimization goodness of fit to a non-metric hypothesis. *Psychometrika* 29: 1–27.

Kruskal, J.B. (1964b). Non-metric multidimensional scaling: A numerical method. *Psychometrika* 29: 115–129.

Kulkarni, S.R. and Paranjape, S.R. (1984). Use of the Andrews' function plot technique to construct control curves for multivariate processes. *Comm. in Statist.— Theor. Meth.* 13: 2511–2533.

Lance, G.N. and Williams, W.T. (1965). Computer programme for monothetic classification ('association analysis'). *Comput. J.* 8: 246–249.

Levy, S. (1985). Lawful roles of facets in social theories. In: Canter, D. (ed.) *Facet theory approaches to social research.* New York: Springer-Verlag.

Mage, D.T. (1982). An objective graphical method for testing normal distributional assumptions using probability plots. *The Amer. Statist.* 36: 166–121.

Magidson, J. (1982). Some common pitfalls in causal analysis of categorical data. *Journal of Marketing Research* xix: 461–471.

Mahon, B.H. (1977). Statistics and decisions: The importance of communication and the power of graphical presentation. *J. Roy. Statist. Assoc.* 140: 298–307.

Mallows, C.L. (1973). Some comments on C_p. *Technometries* 15: 661–675.

Mardia, K.V., Kent, J.T. and Bibby, J.M. (1979). *Multivariate analysis.* London: Academic Press.

Muircheartaigh, C.A. and Payne, C. (ed.) (1977). *The analysis of survey data. Volume* 1. New York: John Wiley.

Murdoch, J. (1979). *Control charts.* London: Macmillian Press.

North, W.R.S. (1980). The Quangle-A modification of the cusum chart. *Applied Statistics* 31: 155–158.

Ord, J.K. (1967). Graphical methods for a class of discrete distributions. *J. Roy. Statist. Assoc. A.* 130: 232–238.

Ord, J.K. (1972). *Families of frequency distributions*: London: Griffen.

Pearson, E.S. and Hartley, H.O. (1966). *Biometrika tables for statisticians.* Vol. 1. Cambridge: Cambridge University Press.

Playfair, W. (1786). *The commercial and political atlas.* London.

Playfair, W. (1801). *The statistical breviary.* London.

Robinson, D.F. and Foulds, L.R. (1980). *Digraphs: Theory and techniques.* New York: Gordon and Breach.

Sampson, F. (1969). Crisis in 'n Cloister (Unpublished Ph.D. dissertation). Cornell University: Dept. of Sociology.

SAS/GRAPH User's Guide (1985). Cary: SAS Institute Inc.

SAS User's Guide: Statistics (1985). Cary: SAS Institute Inc.

Shepard, R.N. (1962). The analysis of proximities: Multidimensional scaling with an unknown distance function. *Psychometrika* 27: 125–140, 219–246.

Schiffman, S.S., Reynolds. M.L. and Young, F.W. (1981). *Introduction to multidimensional scaling.* New York: Academic Press.

Shye, S. (1978). Partial order scalogram analysis, In: Shye, S. (ed.), *Theory construction and data analysis in the behavioral sciences.* San Francisco: Jossey-Bass.

Shye, S. and Amar, R. (1985). Partial-order scalogram analysis by base coordinates and lattice mapping of the items by their scalogram roles. In: Canter, D. (ed.) *Facet theory approaches to social research.* New York: Springer-Verlag.

Sibson, R. and Jardine, N. (1971). *Mathematical Taxonomy.* New York: John Wiley.

Siegel, J.H., Farrell E.J., Goldwyn, R.M. and Friedman H.P. (1972). The surgical implication of physiologic patterns in myocardial infarction shock, *Surgery* 72: 126–141.

SUGI-Supplemental Library User's Guide (1983). Cary: SAS Institute Inc.

Tidmore, F.E. and Turner, D.W. (1977). Clustering with Chernoff-type faces. *Comm. Statist.—Theor. Meth.* 12: 397–408.

Torgerson, W.S. (1958). *Theory and methods of scaling.* New York: John Wiley.

Tukey, J.W. (1977). *Exploratory Data Analysis.* Massachusetts: Addison-Wesley.

Young, G. and Householder, A.S. (1938). Discussion of a set of points in terms of their mutual distances. *Psychometrika* 3: 19–22.

Wainer, H. (1984). How to display data badly. *The Amer. Statist.* 38: 137–147.

Wang, P.C.C. (ed.) (1978): *Graphical representation of multivariate data.* New York: Academic Press.

Ward, J.H. (1963). Hierarchical grouping to optimise an objective function. *J. Amer. Statist. Assoc.* 58: 236–244.

Weisberg, S. (1980). *Applied linear regression.* New York: John Wiley.

Wilk, M.B. and Gnanadesikan, R. (1968). Probability plotting methods for the analysis of data. *Biometrika* 55: 1–17.

Wilk, M.B., Gnanadesikan, R. and Huyett, M.J. (1962). Probability plots for the gamma distribution. *Technometrics* 4: 1–20.

Wishart, D. (1969). Numerical classification for deriving natural classes. *Nature* 221: 97–98.

Index